SpringerBriefs in Electrical and Computer Engineering

For further volumes:
http://www.springer.com/series/10059

Ning Lu • Xuemin (Sherman) Shen

Capacity Analysis of Vehicular Communication Networks

 Springer

Ning Lu
Department of Electrical and Computer
    Engineering
University of Waterloo
Waterloo, Ontario
Canada

Xuemin (Sherman) Shen
Department of Electrical and Computer
    Engineering
University of Waterloo
Waterloo, Ontario
Canada

ISSN 2191-8112   ISSN 2191-8120 (electronic)
ISBN 978-1-4614-8396-0   ISBN 978-1-4614-8397-7 (eBook)
DOI 10.1007/978-1-4614-8397-7
Springer New York Heidelberg Dordrecht London

Library of Congress Control Number: 2013948229

Printed on acid-free paper

Springer is part of Springer Science+Business Media (www.springer.com)

Preface

There have been increasing interest and significant progress in the domain of emerging vehicular communication networks, or simply, VehiculAr NETworks (VANETs). VANETs target the incorporation of wireless communications and informatics technologies into the road transportation system, and thereby facilitate a myriad of attractive applications related to safety (e.g., collision detection and lane changing warning), driving assistance (e.g., online navigation and smart parking), infotainment (e.g., mobile office and music store), and traffic efficiency and management (e.g., real-time traffic notification and electronic tickets). With rapidly evolving concepts of building applications, not only can VANETs make the transportation system safer and more efficient but also revolutionize the in-vehicle experience of passengers.

In this monograph, we focus on the network capacity analysis of VANETs. The problem is of great importance since fundamental guidance on design and deployment of VANETs is very limited. Moreover, unique characteristics of VANETs impose distinguished challenges on such investigations. In Chap. 1, we first give an overview of VANETs and the study of capacity scaling laws. In Chap. 2, we provide a comprehensive survey on the capacity scaling laws in wireless networks. Then, in Chap. 3, we study the unicast capacity considering the socialized mobility model of VANETs. With vehicles communicating based on a two-hop relaying scheme, the unicast capacity bound is derived and can be applied to predict the network throughput in real-world scenarios. In Chap. 4, we investigate the downlink capacity of VANETs in which wireless access infrastructure is deployed to provide pervasive Internet access to vehicles. Different alternatives of wireless access infrastructure are considered. A lower bound of downlink capacity is derived for each type of access infrastructure. We also present a case study based on a perfect city grid to examine the capacity-cost trade-offs of different deployments since the deployment costs of different access infrastructure are highly variable. Finally, we draw conclusions and highlight future research directions in Chap. 5. The implications from this monograph provide valuable guidance on the design and deployment of future VANETs.

We would like to thank Prof. Jon W. Mark, Dr. Tom H. Luan, Miao Wang, Ning Zhang, and Nan Cheng, from Broadband Communications Research Group (BBCR) at the University of Waterloo, and Dr. Fan Bai from General Motor, for their contributions in the presented research works. Special thanks are also due to the staff at Springer Science+Business Media: Susan Lagerstrom-Fife and Courtney Clark, for their help throughout the publication preparation process.

Waterloo, Ontario, Canada   Ning Lu
                                                          Xuemin (Sherman) Shen

Contents

Acronyms

BS	Base Station
B2V	BS-to-Vehicle
CAPEX	CAPital EXpenditures
DSRC	Dedicated Short Range Communications
GPS	Global Positioning System
IPP	Inhomogeneous Poisson Process
ITS	Intelligent Transportation Systems
LOS	Line-Of-Sight
M2M	Mesh-to-Mesh
M2V	Mesh-to-Vehicle
MG	Mesh Gateway
MN	Mesh Node
MPR	Multi-Packet Reception
MR	Mesh Router
OPEX	OPerational EXpenditures
PU	Primary User
QoS	Quality of Service
R2V	RAP-to-vehicle
RAP	Roadside Access Point
RSU	RoadSide Unit
RX	Receiver
SU	Secondary User
TX	Transmitter
UWB	Ultra-WideBand
VANET	VehiculAr NETwork
V2I	Vehicle-to-Infrastructure
V2R	Vehicle-to-Roadside
V2V	Vehicle-to-Vehicle
WMB	Wireless Mesh Backbone

Chapter 1
Introduction

Vehicular networks play an important role in both promoting the development of next generation intelligent transportation systems and offering mobile data services to vehicle users. The capacity scaling laws of vehicular networks characterize the trend of network capacity when vehicle population grows in the network, which represent the fundamental property of vehicular networks and could be applied to predict the network performance and thereby provide valuable guidance on network design and deployment. Despite extensive research in the field of vehicular networking, the network capacity is not well understood. In this chapter, we first overview the vehicular network, and then briefly introduce the research of capacity scaling laws for general wireless networks. Finally, we present the existing works in the capacity study of vehicular networks.

1.1 Overview of Vehicular Networks

There are two major impetuses pushing forward the research of VehiculAr NETworks (VANETs).[1] The first one is the urgent need to improve efficiency and safety of road transportation systems. Due to growing urbanization and environmental pressures, such a need has been increasingly pressing to alleviate transportation problems, including traffic accidents, congestions, and air pollution, among others, especially in the developing world. The second one is the ever-increasing mobile data demand of passengers. As an essential part of our daily life, Internet access is expected anytime and anywhere. People on the move also desire the mobile data service for on-board mobile devices. The emerging VANETs meet these two needs by incorporating wireless communications and informatics technologies into

---

[1] To deemphasize the ad hoc nature of vehicular networks, we redefine the term VANETs, which is traditionally the acronym of vehicular ad hoc networks.

N. Lu and X. Shen, *Capacity Analysis of Vehicular Communication Networks*,
SpringerBriefs in Electrical and Computer Engineering,
DOI 10.1007/978-1-4614-8397-7_1, © The Author(s) 2014

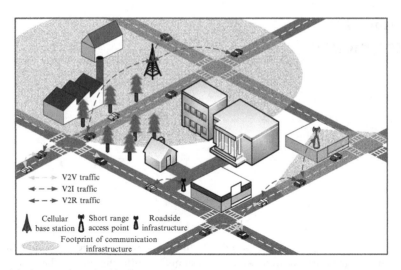

Fig. 1.1 An overview of vehicular networks

the road transportation system, not only enabling the evolution to next generation Intelligent Transportation Systems (ITS), but also catering to the ever-increasing Internet data demand of passengers.

VANETs facilitate a myriad of attractive applications related to safety (e.g., collision detection and lane changing warning), driving assistance (e.g., online navigation and smart parking), infotainment (e.g., mobile office and music store), and traffic efficiency and management (e.g., real-time traffic notification and electronic tickets). Relying on different communication technologies, VANETs basically comprise three types of communications, Vehicle-to-Vehicle (V2V), Vehicle-to-Roadside (V2R), and Vehicle-to-Infrastructure (V2I), as shown in Fig. 1.1.

In the networks of connected vehicles, information generated by the vehicle-borne computer, control system, on-board sensors or passengers can be effectively disseminated among vehicles in proximity by means of V2V communications. The realization of V2V communications can resort to Dedicated Short Range Communications (DSRC) [1], tailored from the Wi-Fi technology. Without the assistance of any built infrastructure, a variety of active road safety applications (e.g., collision detection, lane changing warning, and cooperative merging) [2] and a small number of infotainment applications (e.g., interactive gaming, and file and other valuable information sharing) [3–5] are enabled by using V2V communications.

Providing high-rate Internet access for vehicles can be envisioned not only to cater to the ever-increasing Internet data demand of passengers, but also enrich safety-related applications, such as online diagnosis [6], and intelligent anti-theft and tracking [7], in which the servers can be on Internet cloud. A recent automotive executive survey [8] further reveals that Internet access is predicted to become a standard feature of motor vehicles. Communications between vehicles and Internet-access infrastructure are referred to as V2I communications. The most practical and seamless way to provide Internet connectivity to vehicles is through the use of wireless wide area network [9], such as off-the-shelf 3G or 4G-LTE cellular networks. Due to the relatively high cost of cellular services, people may like to use much cheaper access technologies, such as the "grassroots" Wi-Fi access point. By equipping the Wi-Fi radio, vehicles can access the Internet when they are moving on the road. Such kind of access network is often named drive-thru Internet in the literature [10]. The problem of using Wi-Fi access points is that one has to tolerate intermittent connectivity, as mentioned in a real-world measurement study of the drive-thru Internet [11]. Another possible solution to provide Internet access for vehicles is through the use of a fixed wireless mesh backbone [12], which consists of wirelessly connected routers and mesh gateways to the Internet. It is expected that such a mesh structure can be a compromise between high cost and poor connectivity. To extend the service range of communication infrastructures, V2V communications may be utilized to relay the V2I traffic, as shown in Fig. 1.1, so that it is possible to route packets from/to the infrastructure through multiple hops.

Leveraging short range communications (e.g., DSRC) or sensor technologies can yield V2R communications, in which roadside infrastructures do not serve as Internet access as communication infrastructures in V2I communications. By mounting the radio of V2R communications, vehicles are allowed to communicate with roadside or ITS infrastructures, such as traffic lights, street signs, roadside sensors, etc., to avoid or mitigate the effects of road accidents, and to enable the efficient management of ITS. Moreover, roadside infrastructures can also be commercial content providers, such as the roadside unit broadcasting flyers of superstores [13]. If we extend the scope of "roadside", communications between vehicles and satellites fall into the category of V2R communications. One well-known example is Global Positioning System (GPS) which has been highly commercialized. Based on these three types of communications and with rapidly evolving concepts of building applications, not only can VANETs make the transportation system safer and more efficient, but they can also revolutionize the in-vehicle experience of the passengers with media-rich infotainment.

Fascinated by the concepts and visions of VANETs, the academia, industry, and government institutions have initiated numerous activities. An overview of the current and past major programs of the ITS and projects in the USA, Japan, and Europe can be seen in [14]. The standards and standardization process of VANETs are given in [15, 16]. There have been tons of research works produced in the past decade to speed up the development of VANETs, including comprehensive surveys (e.g., [14] and [17]).

1.2 Capacity Scaling Laws

Wireless networks have received a myriad of research attentions over the past decades, including medium access control, routing, security, cooperation, and energy-efficiency, among others. Despite significant advances in the field of wireless networking, a fundamental question remains open: how much information can a wireless network transfer? To answer this question, we should resort to the study of network capacity which is a central concept in the field of network information theory [18]. Intuitively, if the capacity of a wireless network is known, the network limit of information transfer would be obtained. Moreover, having such knowledge would shed light on what the appropriate architectures and protocols were for operating wireless networks. Although significant efforts have been put on the investigation of network capacity, developing a general theory of such fundamental limit for wireless networks is a long standing open problem [19]. In [20], Claude Shannon successfully determined the maximum achievable rate, called the capacity, for a point-to-point communication channel, below which the reliable communication can be implemented while above which the reliable communication is impossible. However, general wireless networks with sources and destinations sharing channel resources are much more complex, making the quest for fundamental limits of wireless networks a formidable task. For example, even for a simple-looking three-node relay channel [21], the exact capacity still has yet to be determined.

As a retreat when exact fundamental limits are out of reach, capacity scaling laws, first studied by Gupta and Kumar in [22], characterize the trend of node throughput behavior when the network size increases. The most salient feature of capacity scaling studies is to depict the capacity as a function of the number of nodes in the network, without distractions from minor details of network protocol. This approach is quite different from that of studying network information theory, which is to determine exact capacity region of wireless networks. The seminal work [22] not only provides an alternative and tractable way to study the network capacity, but also obtains insightful capacity results. Great efforts have been made thereafter to derive capacity scaling laws for different paradigms of wireless networks. Scaling laws for network delay and its trade-off with capacity have also been investigated. The study of scaling laws can lead to a better understanding of intrinsic properties of wireless networks and theoretical guidance on network design and deployment [23]. Moreover, the results could also be applied to predict network performance, especially for the large-scale networks [24]. We provide the following illustration. We consider to deploy a large-scale sensor networks for a certain geographic area. Capacity scaling laws show that the network scales poorly when the number of sensors grows, i.e., the throughput of each sensor would decrease. In order to enhance the throughput capacity, we may need to adopt some advanced technologies, such as directional antennas and network coding. However, scaling laws show that exploiting network coding cannot change the trend of throughput capacity; whereas exploiting directional antennas can introduce capacity gains (refer to Sect. 2.3.1). Furthermore, suppose we have deployed a sensor network of 100 sensors with directional antennas. Typically we can obtain the throughput performance (denoted by λ_A) of the network through real measurement. If we need to extend the network to a larger one of 1,000 sensors, with the same network settings, by capacity scaling results (denoted by $f(N)$), we are able to have a rough idea that how much throughput (denoted by λ_B) can be supported by the network that we are going to deploy, i.e., $\lambda_B = \lambda_A \cdot f(1000)/f(100)$. We survey the literature in this area in Chap. 2.

1.3 Capacity of Vehicular Networks

The capacity scaling of VANETs is desirable since unlike generic mobile ad hoc networks, VANETs present unique characteristics, which impose distinguished challenges on networking. (i) *Large scale*: the VANET is an extremely large-scale mobile network, which is deployed in a large geographic area with a great amount of vehicles and roadside units; (ii) *Cars on the road:* the movement of vehicles should follow certain street pattern, different from generic mobile ad hoc networks in which nodes typically move in a free space; (iii) *Cars on wheels:* the vehicle mobility is related to the road traffic environment and the social life of the driver; (iv) *Spatio-temporal variations:* there are spatio-temporal variations of vehicle density and link quality due to vehicle mobility and unstable wireless channels, respectively; and (v) *Diversified applications:* applications of VANETs are of a large variety and with different quality of service (QoS) requirements. All these features dramatically complicate studies of scaling laws.

There have been a few efforts to investigate the capacity of VANETs. Pishro-Ni et al. [25] initiated the study of capacity scaling for vehicular networks with an emphasis on the impact of road geometry on the network capacity. Nekoui et al. [26] specially developed a novel notion of capacity for safety applications, which is called *Distance-Limited Capacity*. That is the capacity of VANETs when a pair of vehicles can only communicate if the two vehicles reside in a certain distance of each other. Both [25] and [26] showed that the road geometry has an important role in the capacity of vehicular networks. As the demand of public information dissemination is high in VANETs, multicast flows, in which one source is associated with a set of destinations, may be viable to be deployed for practical applications. In [27], Zhang et al. analyzed multicast capacity of hybrid VANETs, in which base stations are deployed to support communications between vehicles. It was assumed that each vehicle is equipped with a directional antenna. By respectively applying the one-dimensional and two-dimensional i.i.d mobility model (refer to Sect. 2.3.1) to vehicles, they derived bounds of the multicast throughput capacity under certain end-to-end delay constraint. In [28], Wang et al. studied the uplink capacity of hybrid VANETs, where each vehicle, following random way-point mobility, is required to send packets to regularly placed sink roadside units (RSUs). The basic routing strategy adopted in [28] is to distribute source packets to as many RSUs as possible to increase concurrent uploading opportunities.

One of the limitations of [27] and [28] is that the specific mobility features of vehicles are not fully considered. The i.i.d mobility is not practical for vehicular scenarios. Moreover, the assumption that vehicles are uniformly distributed in the network is also unrealistic. In urban areas, vehicle densities in different regions may be highly diverse. Inhomogeneous vehicle densities were considered in [29], which investigates the throughput capacity of social-proximity VANETs. We will elaborate this work in Chap. 3. Wireless access infrastructure, such as Wi-Fi access points and cellular base stations, plays a vital role in providing pervasive Internet access to vehicles. Reference [30] analyzes the downlink capacity of vehicles, i.e., the maximum average downlink rate achieved *uniformly* by vehicles, for each type of access infrastructure considered, and investigates the capacity-cost tradeoffs for access infrastructures since the deployment costs of different access infrastructure are highly variable. We will elaborate this work in Chap. 4.

Chapter 2
Capacity Scaling Laws of Wireless Networks

The capacity scaling law of wireless networks has been considered as one of the most fundamental issues. In this chapter, we aim at providing a comprehensive overview of the development in the area of scaling laws for throughput capacity in wireless networks. We begin with background information on the notion of throughput capacity of random networks. Based on the benchmark random network model, we then elaborate the advanced strategies adopted to improve the throughput capacity, and other factors that affect the scaling laws. We also present the fundamental tradeoffs between throughput capacity, delay, and mobility. Finally, the capacity for hybrid wireless networks are surveyed, in which there are at least two types of nodes functioning differently, e.g., normal nodes and infrastructure nodes.

2.1 Preliminaries: Milestone of Throughput Capacity Scaling

Capacity scaling laws offer fundamental understanding on how per-node capacity scales in an asymptotically large network. The line of investigation began with [22], where Gupta and Kumar introduced two new notions of network capacity: *transport capacity* and *throughput capacity*. In this monograph, we focus on the throughput capacity. We first introduce the notion of throughput capacity and the capacity result for random networks, as preliminaries for reading the remaining sections.

2.1.1 Notion of Throughput Capacity

Let N denote the number of nodes in a network. The per-node throughput of the network, denoted by $\lambda(N)$, is the average transmission rate, measured in bits or packets per unit time, that can be supported uniformly for each node to its destination in the network. A per-node throughput of $\lambda(N)$ bits per second

N. Lu and X. Shen, *Capacity Analysis of Vehicular Communication Networks*,
SpringerBriefs in Electrical and Computer Engineering,
DOI 10.1007/978-1-4614-8397-7_2, © The Author(s) 2014

is said to be *feasible* if there exists a spatial and temporal scheme for scheduling transmissions, such that each node can send $\lambda(N)$ bits per second on average to its destination node. The throughput capacity of the network is said of order $\Theta(f(N))$[1] bits per second if there are deterministic constants $c_1 > 0$ and $c_2 < \infty$ such that

$$\lim_{N \to \infty} \mathbf{Pr}\big(\lambda(N) = c_1 f(N) \text{ is feasible}\big) = 1$$

$$\liminf_{N \to \infty} \mathbf{Pr}\big(\lambda(N) = c_2 f(N) \text{ is feasible}\big) < 1.$$

Therefore, vanishingly small probabilities are allowed for in such definition of "throughput capacity" when considering the randomness involved in the network, such as the location and the destination of each node. Note that the notion of throughput capacity is different from the information-theoretic capacity notion that describes the exact region of simultaneous rates of communications from many senders to many receivers in the presence of interference and noise [31].

2.1.2 Random Networks

A wireless random network consisting of N identical immobile nodes randomly located in a disk of unit area in the plane and operating under a *multi-hop* fashion of information transfer, is shown in Fig. 2.1 [22]. Each node having a randomly chosen destination is capable of transmitting at W bits per second over a common wireless channel. The requirements for successful transmission are described as per two interference models: (i) the Protocol Model, which is a binary model, i.e., the transmission is successful if there is enough spatial separation from simultaneous transmissions of other nodes otherwise fails; and (ii) the Physical Model, based on signal-to-interference ratio requirements. In such static random ad hoc network, all the nodes are assumed to be homogeneous, i.e., all transmissions employ the same range or power, and wish to transmit at a common rate.

---

[1] Since studies of throughput capacity focus on the scaling behavior instead of a specific value, the order notation is involved to describe how capacity scales with the number of nodes N. Specifically, the following Knuth's notation is used throughout all the papers on scaling laws: given nonnegative functions $f_1(n)$ and $f_2(n)$, $f_1(n) = O(f_2(n))$ means $f_1(n)$ is asymptotically upper bounded by $f_2(n)$; $f_1(n) = \Omega(f_2(n))$ means $f_1(n)$ is asymptotically lower bounded by $f_2(n)$; and $f_1(n) = \Theta(f_2(n))$ means $f_1(n)$ is asymptotically tight bounded by $f_2(n)$; $f_1(n) = \omega(f_2(n))$ means $f_1(n)$ is asymptotically dominant with respect to $f_2(n)$; $f_1(n) = o(f_2(n))$ means $f_1(n)$ is asymptotically negligible with respect to $f_2(n)$.

Fig. 2.1 A static ad hoc
network in a unit disk

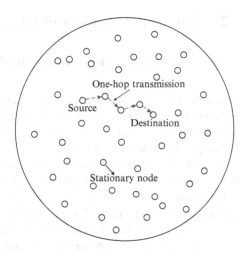

2.1.3 Throughput Capacity of Random Networks

For random networks, the order of the throughput capacity is $\lambda(N) = \Theta(\frac{W}{\sqrt{N \log N}})$
under Protocol Model (see main result 3 in [22]); while under Physical Model,
the throughput capacity is given by $\Theta(\frac{W}{\sqrt{N \log N}}) \leq \lambda(N) < \Theta(\frac{W}{\sqrt{N}})$ (see main
result 4 in [22]). An explanation of the results is as follows. For Protocol Model,
the lower bound and upper bound are of the same order such that there exists a sharp
order estimation of the throughput capacity; for Physical Model, a throughput of
order $\Theta(\frac{W}{\sqrt{N \log N}})$ is feasible, while $\Theta(\frac{W}{\sqrt{N}})$ is not.

The throughput capacity is studied asymptotically, i.e., the population of nodes
continuously grows. The scaling result for random networks is pessimistic because
the per-node throughput tends to zero similar to $\frac{1}{\sqrt{N \log N}}$ as the population of nodes
goes to infinity, which indicates that static ad hoc networks are not feasible to
scale to a large size. What causes those discouraging results? The fundamental
reason is that every node in the network needs to share the channel resources
or certain geographic area with other nodes in proximity, which constricts the
capacity. Specifically, concurrent wireless transmissions in a wireless network limit
its throughput capacity, because they create mutual interference so that nodes cannot
communicate as that in the wireline network where much less mutual interference
exists. This interpretation also demonstrates how desirable it is to mitigate the
mutual interference in wireless communications, although it is very challenging.

2.2 Throughput Capacity of Ad Hoc Networks

2.2.1 Strategies to Improve Throughput Capacity

One natural question is if it is possible to improve throughput capacity of random networks by employing any advanced techniques or sophisticated strategies. After significant progress that has been made to further the investigation on throughput capacity scaling, the answer is positive.

First of all, by allowing both long-distance and short-distance transmissions, the throughput capacity can be improved slightly to $\Theta(\frac{1}{\sqrt{N}})$ [32]. The scheme constructed to achieve this throughput relies on multi-hop transmission, pairwise coding and decoding at each hop, and a time-division multiple access. The gain of throughput capacity can also be achieved by employing directional antennas. Yi et al. in [33] considered different beamform patterns, and showed that the throughput capacity can be achieved with a gain of $\frac{4\pi^2}{\alpha\beta}$ using directional transmission and reception, where α and β are antenna parameters. Peraki et al. in [34] further revealed that the maximum capacity gain is $\Theta((\log N)^2)$ by using directional antennas at the transmitters and receivers, corresponding to a throughput of $\Theta((\log N)^{3/2}/\sqrt{N})$. If nodes have multi-packet reception (MPR) capabilities, i.e., a receiver is capable of correctly decoding multiple packets transmitted concurrently from different transmitters, the capacity gain can also be achieved. Sadjadpour et al. in [35] showed that with MPR, the throughput capacity of random ad hoc networks can be improved at least by an order of $\Theta(\log N)$ and $\Theta((\log N)^{\frac{\alpha-2}{2\alpha}})$ under Protocol Model and Physical Model, respectively, where α is the path loss exponent in the Physical Model. Similar research efforts applying MPR can be found in [36–38].

By means of long-range multiple-input multiple-output (MIMO) communications with local cooperations as proposed in [39], significant improvement of throughput capacity scaling in random networks is attainable [40], i.e., almost constant per-node throughput of $\Theta(n^{-\epsilon})$ on average is achievable, where $\epsilon > 0$ can be arbitrarily small. This yields an aggregate throughput $(N\lambda(N))$ of $\Theta(N^{1-\epsilon})$ for the whole network, indicating almost linear capacity scaling in N. $\epsilon = \Theta(\frac{1}{\sqrt{\log N}})$ was explicitly obtained later in [41] and [42]. However, the capacity gain is at the cost of increased system complexity due to the intelligent hierarchical cooperation among nodes. Regardless the complexity of the constructed strategy, the result in [40] is inspiring but still controversial. Franceschetti et al. in [43] claimed that a throughput higher than $O((\log N)^2/\sqrt{N})$ cannot be achieved because of degrees of freedom limitation which is a result of laws of physics. Artificial assumptions and models lead to the impossible linear capacity scaling in [40]. While using Maxwell's equations without any artificial assumptions, Lee et al. in [44] established the capacity scaling laws for the line-of-sight (LOS) environments, which show that a linear scaling of the aggregate throughput is indeed possible for static random networks. Thus, the conflict between [40] and [43] was resolved. It is worth noting that even if such physical limits in [43] do exist and sophisticated strategies like the hierarchical cooperation cannot further improve the per-node throughput ($\Theta(1/\sqrt{N})$) in the scaling limit sense, these strategies generally could be considerably beneficial in networks of any finite size. An example is the physical-layer network coding. In [45], it was shown that although the physical-layer network coding scheme does not change the scaling law, it improves throughput performance of the network in the sense by enlarging the constant component of the scaling result. The similar studies applying the network coding schemes can be found in [46–49].

Since the above research works are based on the assumption that the network is bandwidth-constrained, i.e., each node is only capable of transmitting at W bits per second, it is interesting to consider a scenario where each node has power constraint but can utilize unlimited bandwidth. Hence, there have been a few research efforts which focus on the ultra-wideband (UWB) techniques. In [50], Negi and Rajeswaran showed that under the limiting case $W \rightarrow \infty$, the throughput capacity is lower bounded by $\Omega\left(P_0\sqrt{N^{\alpha-1}/(\log N)^{\alpha+1}}\right)$ and upper bounded by $O(P_0(\sqrt{N \log N})^{\alpha-1})$, where α is the path loss exponent and P_0 is the maximum transmission power. The gap between the upper bound and the lower bound was closed by Tang and Hua in [51]. They showed that the throughput capacity of a UWB power-constrained ad hoc network is given by $\Theta(P_0(\sqrt{N/\log N})^{\alpha-1})$. A better result was obtained in [52] that the throughput capacity scales as $\Theta(P_0 N^{(\alpha-1)/2})$.

Without leveraging aforementioned advanced techniques in the static random network, what if nodes move? The effect of nodal mobility on throughput capacity scaling was first investigated by Grossglauser and Tse in [53]. By applying an i.i.d mobility model (see Sect. 2.3) to each node, they have shown that the per-node throughput of the mobile ad hoc network could remain constant, i.e., $\Theta(1)$, by using a **two-hop relaying** scheme (see Fig. 2.2c) and allowing finite but arbitrary delay. This result provides an interesting implication that dramatic gains in network capacity are possible when mobility is considered so that the nodes can exploit mobile relays to carry packets to distant nodes. Compared with such *store-carry-and-forward* communication paradigm, in the absence of mobility, direct transmission (see Fig. 2.2a) between distant nodes causes too much interference, or equivalently requires a large spatial area, so that the number of concurrent transmissions are reduced; on the other hand, if the network only allows the communication between nearest neighbors (see Fig. 2.2b), most of the packets will be delivered through multiple hops, resulting in the decrease in throughput capacity as well. Inspired by the promising result in [53], extensive works have been done to investigate capacity scaling in mobile ad hoc networks. In [54], Diggavi et al. have shown that even one-dimensional mobility benefits capacity scaling. In [55], Syed Ali Jafa explored the capacity of high mobile ad hoc networks in the presence of channel uncertainty, and has shown that high mobility introduces rapid channel fluctuations and hence limits the capacity of wireless networks.

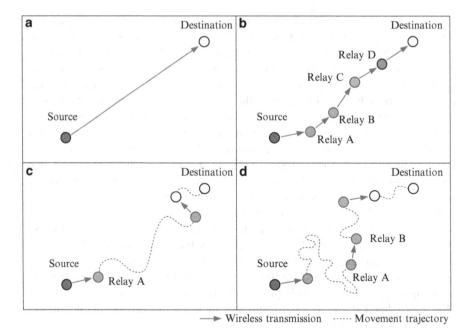

Fig. 2.2 An illustration of packet transmission strategies. (**a**) Direct transmission. (**b**) Multi-hop relying in a static ad hoc network. (**c**) Two-hop relaying in a mobile ad hoc network. (**d**) Multi-hop relying in a mobile ad hoc network

2.2.2 *Other Factors Affecting Scaling Laws*

The random network considered in [22] is a benchmark network model, in which nodes have basic communication capabilities (i.e., simple coding and decoding strategies implemented on the single radio), and the traffic model (symmetric unicast) and interference model (Protocol Model or Physical Model) are simplified. Besides the strategies mentioned in Sect. 2.2.1 to improve throughput capacity, significant research efforts have been made to study the impact of different modeling factors on capacity scaling laws.

Multi-channel multi-interface: In [22], it has been shown that with a single radio mounted on each vehicle, splitting the total bandwidth W into multiple sub-channels does not change the order of throughput capacity of random networks. However, in practice, a communication device may have multiple radio interface operating on one or different channels. What if each node is equipped with multiple radio interfaces? In [56] and [57], Kyasanur and Vaidya derived the capacity scaling laws for a general multi-channel networks with $m \leq c$, where c is the number of channels, and m is the number of interface per node. It was shown that different ratios between c and m yield different capacity bounds, and in general, the network capacity is reduced except when c is upper bounded by $O(\log N)$. Kodialam and Nandagopal [58] also provided capacity trends for multi-channel multi-interface wireless mesh networks by considering channel assignment and scheduling.

Channel model: Most research efforts follow either Protocol Model (governed by geometry) or Physical Model (governed by path loss), which only characterize the deterministic behavior of wireless channel connection. To consider the channel randomness which is more realistic, several works have been done assuming different channel models. The impact of channel fading on capacity scaling was studied by Toumpis and Goldsmith [59]. They showed that a throughput of $\Theta(\frac{1}{\sqrt{N(\log N)^3}})$ is feasible under a general model of fading for static random networks. The Rayleigh fading was considered by Ebrahimi et al. [60] and the lower and upper bounds of throughput capacity were derived. The random connection model was considered in [61] and [62], i.e., the signal strengths of the connections between nodes are independent from each other and follow a common distribution. In [63], Gowaikar and Hassibi considered a *hybrid* channel connection model: for a short distance between transmitter and receiver, the channel strengths are governed by the random connection model; while for a long range, the channel strengths are governed by a Rayleigh distribution. They showed that a throughput of $\Theta(\frac{1}{(\log N)^4})$ is achievable. The lower bound on the capacity of wireless erasure networks was reported by Jaber and Andrews [64], in which an erasure channel model is considered, i.e., each channel is associated with an erasure probability. Such a channel model incorporates erasure events which may correspond to packet drops or temporary outages when transmission is undergoing.

Network topology: The shape of geographic area where the network is deployed has a significant impact on capacity scaling laws. Hu et al. [65] investigated the effect of various geometries, including the strip, triangle, and three-dimensional cube. The main implication from [65] is that the symmetry of the network shape plays an important role. In other words, a high throughput capacity can be achieved if the network is symmetric. In [66], Li et al. respectively derived the capacity bound for the three-dimensional network with regularly and randomly deployed nodes.

Traffic pattern: Besides symmetric unicast, i.e., each node is only the source of one unicast flow and the destination of another, dissemination of information in other fashions has been extensively studied in the literature. The broadcast capacity is reported in [67–69], which is the maximum per-node throughput of successfully delivered broadcast packets. For each broadcast packet, it is successfully delivered if all nodes in the network other than the source receive the packet correctly in a finite time. The multicast capacity has been widely investigated [70–76] considering different network settings. By employing multicast, each packet is disseminated to a subset of $N - 1$ nodes which are interested in the common information from the source. Nie [77] reported a short survey on multicast capacity scaling. A unifying study was provided by Wang et al. [78], in which how information is disseminated is generally modeled by the (n, m, k)-casting. In this particular context, n denotes the population of nodes, and m and k denote the number of intended recipients of a source packet and the number of successful recipients, respectively. For unicast, $m = k = 1$; for multicast, $k \leq m < n$; and for broadcast, $k \leq m = n - 1$. The capacity bounds were established in [78] for each type of traffic pattern.

2.3 Fundamental Tradeoffs: Capacity, Delay, and Mobility

For network performance, capacity is not the only metric. From applications point of view, network delay (its average, maximum, or distribution) is also an important design aspect [23]. In [53], it has been shown that striking performance gains in throughput capacity are achievable in mobile ad hoc networks, however at the expense of enlarged delay. With the same time scale of node mobility, the delay is incurred by the **movements** of the relay (transmitter) and the destination (receiver) since they have to be geographically close enough for transmission, as shown in Fig. 2.2c. Basically, there are two ways to transfer a piece of information from the source to the destination: wireless transmission and node movement. Since wireless transmission is typically at a much smaller time scale, the time spent on the relay movements towards the destination contributes to the major part of the delay. There is a tradeoff between capacity and delay: if an increase in throughput is desired, we should reduce the distance of wireless transmission to allow more concurrent transmissions in the network; while if a decrease in delay is desired, we should reduce the distance of relay movement towards the destination. However, it is impossible to reduce both distances simultaneously given a fixed distance between the source and the destination. Furthermore, intuitively, different mobility models may incur different delays, because the node movement pattern determines the time spent on the relay movements. For example, if a node always wanders around (see Relay A in Fig. 2.2d), it is very difficult for the node to move a long distance in one direction. Therefore, quite straightforward research studies have been done to understand the tradeoffs between capacity and delay of wireless networks under a variety of mobility models.

2.3.1 Mobility Models

The type of node mobility studied includes the i.i.d mobility, random walk model, random way-point model, Brownian motion, and Lévy mobility. Besides, there are two more general mobility models defined in [79] to study the relationship between delay and throughput capacity from a global perspective.

- i.i.d Mobility Model: In time-slotted system, at each time slot, each node selects a new position independently and identically distributed over all positions in the network. The position distributions of the nodes are independent between time slots. The i.i.d mobility is also referred to as the reshuffling model [80]. Depicting an extreme mobility, the i.i.d mobility model is unrealistic but analytically tractable.

- Random Walk Model: Random walk can be described by Markovian dynamics from i.i.d mobility and is often considered symmetric, i.e., nodes select new positions for next time slot equally likely from the set of current neighboring positions.

- Random Way-Point Model: In random way-point model, at each time slot, the mobile node chooses a random destination in the network and moves toward it at a random speed. The node pauses for some random time after reaching the destination, and then repeats this process.

- Brownian Motion: Brownian motion is like the motion conducted by a small particle totally immersed in a liquid or gas. Brownian mobility has a strong connection with random walk model and is a limiting case when taking smaller and smaller steps in smaller and smaller time intervals in symmetric random walk [81].

- Lévy Walk and Lévy Flight: Lévy mobility is a special type of random walk in which the distribution of flights, i.e., step-lengths, is heavy tailed. In other words, the trajectory of Lévy mobility contains many short flights and an exponentially small number of long flights. The difference between Lévy walk and Lévy flight is that the former has constant flight speed and the latter has constant flight time [82]. It is reported that Lévy mobility has certain statistical similarity to human mobility and some animals' hunting patterns [83].

- Hybrid Random Walk Models: A family of hybrid random walk models is considered in [79] and characterized by a single parameter $\beta \in [0, \frac{1}{2}]$. The unit area of the network is divided into $N^{2\beta}$ equal-sized squares, each of which is further divided into $N^{1-2\beta}$ equal-sized sub-squares. At the beginning of each time slot, each node jumps from its current sub-square to a random sub-square of one uniformly selected neighboring square, as shown in Fig. 2.3a. It can be seen that the model turns to the i.i.d mobility model and the random walk model when $\beta = 0$ and $\beta = 1/2$, respectively.

- Discrete Random Direction Models: A family of discrete random direction models is also considered in [79] and characterized by a single parameter $\alpha \in [0, \frac{1}{2}]$. The unit area of the network is divided into $N^{2\alpha}$ squares with equal area. The movement of each node is of the following pattern: at the beginning of each time slot, the node jumps from its current square to a uniformly selected neighboring square; and during the time slot, the node moves from a start position to an end position at a certain velocity, as shown in Fig. 2.3b. The two positions are uniformly selected from all the positions in the square. It can be seen that the above mobility model turns to the random way-point model and the discrete Brownian motion when $\alpha = 0$ and $\alpha = 1/2$, respectively.

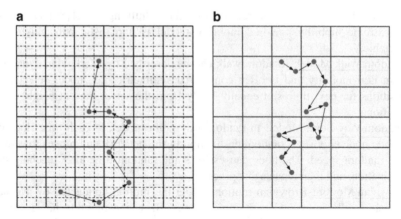

Fig. 2.3 An example of trajectories of hybrid random walk (**a**) and random direction model (**b**), respectively

2.3.2   Tradeoffs Between Throughput Capacity and Delay

The throughput capacity and delay under the i.i.d. mobility model were reported by Neely and Modiano [84] for a cell-partitioned ad hoc network. They found that a general delay-throughput tradeoff can be established: the ratio of delay and throughput is at least $O(N)$ under different scheduling policies (i.e., two-hop or multi-hop relaying) with or without packet redundancy.[2] The optimality of delay-capacity tradeoffs under i.i.d. mobility model was studied in [85]. Different time scales of nodal mobility are taken into consideration: fast mobility, only allowing one-hop transmissions during a time slot after which the topology changes; and slow mobility, allowing packets to be delivered through multiple hops during a time slot since the mobility of nodes is much slower than packet delivery time. It was shown that under i.i.d. fast mobility, a per-node capacity is $O(\sqrt{\mathscr{D}/N})$ given a delay constraint $\mathscr{D}$; while a per-node capacity is $O(\sqrt[3]{\mathscr{D}/N})$ under i.i.d. slow mobility, which is a tighter bound than $O(\sqrt[3]{\mathscr{D}/N}\log N)$ obtained in [86].

In [87], El Gamal et al. studied the throughput and delay under random walk model. It was shown that the ratio of delay and throughput is $\Theta(N)$ for throughput of $O(\frac{1}{\sqrt{N}\log N})$, while the delay remains $\Theta(N\log N)$ for almost any throughput of a higher order, indicating an unsmooth tradeoff under random walk model. Similar insights can be obtained for Brownian motion. In [88], Lin et al. first derived a lower bound of $\Omega(\log N/\sigma^2)$ for average delay associated with capacity of $\Theta(1)$ by using the two-hop relaying scheme proposed in [53], where σ^2 is related to the Brownian mobility model. More importantly, they demonstrated that it is impossible to reduce

---

[2]Redundancy of the packet means extra copies of the original packet, which are issued by the source node.

Table 2.1 Summary of capacity-delay tradeoffs for random ad hoc networks[a]

	Two-hop delay	Critical delay	Any tradeoff?
i.i.d Mobility	$\Theta(N \log N)$ [87]	$\Theta(1)$ [86]	Yes
Random walk	$\Theta(N \log N)$ [87]	$\Theta(N \log N)$ [87]	No
Random way-point	$\Theta(N)^{b}$[89]	$\approx \Theta(\sqrt{N})$ [79]	Yes
Brownian motion	$\approx \Theta(N)$ [79]	$\Theta(N)$ [82]	No
Discrete random direction (α)	$\Theta(N)$ [79]	$\approx \Theta(N^{\alpha+0.5})$ [79]	Yes
Hybrid random walk (β)	$\Theta(N)$ [79]	$\Theta(N^{2\beta} \log \log N)$ [79]	Yes
Lévy walk (γ)	$\approx \Theta(N)$ [82]	$\Theta(N^{\frac{1}{2}})$ for $0 < \gamma < 1$;	
		$\Theta(N^{\frac{\gamma}{2}})$ for $1 \le \gamma < 2$ [82]	Yes
Lévy flight (γ)	$\approx \Theta(N)$ [82]	$\Theta(N^{\frac{\gamma}{2}})$ [82]	Yes

[a] The result is for the case in which the velocity does not scale with the network size
[b] $\alpha \in (0, 0.5)$, $\beta \in (0, 0.5)$ and $\gamma \in (0, 2]$

a large amount of delay without dropping the throughput to $O(\frac{1}{\sqrt{N}})$. From [87] and [88], significant increase in delay cannot be circumvented if a larger throughput than $\Theta(\frac{1}{\sqrt{N}})^{3}$ is desired by using random walk mobility or Brownian motion. Without showing any tradeoff, Sharma and Mazumda [89] analyzed the average delay of the two-hop relaying scheme in a network of N nodes following random way-point mobility.

To further investigate the impact of nodal mobility on throughput capacity and delay, Sharma et al. [79] proposed two general classes of mobility models, i.e., hybrid random walk models and discrete random direction models, incorporating mobility models aforementioned in [84, 87, 88]. The objective of this systematical study is to inquiry how much delay the network has to bear to achieve a per-node capacity better than $\Theta(\frac{1}{\sqrt{N}})$ under different mobility models, resulting in the notion of *critical delay*. Considering that the worst performance in network delay is incurred by the two-hop relaying scheme (*two-hop delay*), however, with an optimal throughput, the room left for tradeoff is actually determined by these two important delays. In [79], it was shown that tradeoffs are negligible under random walk model and Brownian motion, as also shown in [87] and [88], respectively; However, the tradeoff between delay and capacity is quite smooth under i.i.d. mobility and random way-point model. In [82], Lee et al. studied the delay-capacity tradeoffs under Lévy mobility. By using the limiting features of the joint spatio-temporal probability density functions of Lévy models, they derived the critical delay under Lévy walk and Lévy fight, respectively. It was shown that smooth tradeoffs can be obtained and are determined by the distribution parameter related to Lévy mobility. A summary of delay-capacity tradeoffs for random ad hoc networks is given in Table 2.1. Figure 2.4 also shows delay-capacity tradeoff regions under different mobility models.

---

[3]The throughput is achievable in static random ad hoc networks.

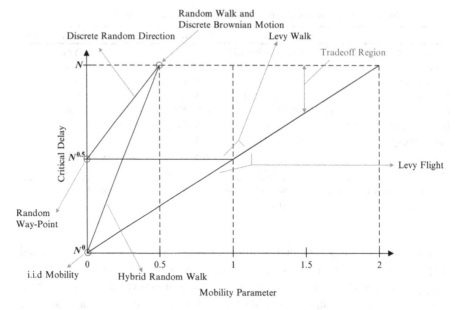

Fig. 2.4 Tradeoff regions for a particular mobility parameter under different mobility models

2.3.3 Impact of Restricted and Correlated Mobility

The mobility models considered in aforementioned delay-capacity studies rely on the following assumptions: (i) the mobility pattern of each node is identical; (ii) following certain ergodic mobility process, each node can visit the entire network area equally likely; and (iii) the movements of different nodes are independent. There have been several efforts made by follow-up investigations to relax these assumptions and then find the impact of restricted and correlated mobility on delay and throughput performance in ad hoc networks.

Restricted Mobility: By noticing that nodes often spend most of the time in proximity of a few preferred places within a localized area, some researchers have studied the throughput and delay under the restricted node mobility, which is more realistic to characterize mobility traces of humans, animals, and vehicles. Li et al. [90] investigated the impact of a restricted mobility model on throughput and delay of a cell-partitioned network. They found that smooth throughput-delay tradeoffs in mobile ad hoc networks can be obtained by controlling the mobility pattern of nodes. Unlike the network in [90] showing homogeneous node density, Garetto et al. have done a series of research [91–95] on the network with heterogeneous node density under restricted mobility model. The capacity scaling of a class of mobile ad hoc networks which show spatial inhomogeneities by considering a cluster mobility model was analyzed in [93] and [94]. In [95], Garetto and Leonardi demonstrated that the delay-throughput tradeoffs can be improved by restricting the node mobility. They considered a restricted mobility that the node moves around a fixed home-point according to a Markov process, and the stationary distribution of the node location decays as a power law of exponent δ with the distance from the home-point. They showed that it is possible to exploit node heterogeneity under a restricted mobility model to achieve $\Theta(1/\log^2(N))$ throughput capacity and $O(\log^4(N))$ delay by using a sophisticated bisection routing scheme.

Correlated Mobility: Instead of exploring the full range of possible capacity-delay trade-offs, Ciullo et al. [80] studied the impact of correlated mobility on performance of delay and throughput capacity. They considered a mobility model in which nodes in the network are grouped and each group, occupying a disc area, moves following i.i.d mobility. Although each node visits uniformly the entire network, movements of different nodes belonging to the same group are not independent. It was shown that the correlated mobility pattern has significant impact on asymptotic network performance and it is possible to achieve better delay and throughput performance than that shown in [85].

2.3.4   Capacity and Delay Scaling without Exploiting Mobility

In [87], El Gamal et al. established delay-capacity tradeoffs for static ad hoc networks. It was shown that the tradeoff when applying multi-hop schemes is given by $\mathscr{D} = \Theta(N\lambda)$, where λ and $\mathscr{D}$ are respectively the throughput and delay. Following [40], throughput and delay tradeoff by means of hierarchical cooperation has been studied in [96], showing that $\mathscr{D} = N \log^2(N)\lambda$ for λ between $\Theta(\frac{1}{\sqrt{N}\log N})$ and $\Theta(\frac{1}{\log N})$. To serve delay sensitive traffic, Comaniciu and Poor [97] reported the delay-constrained capacity scaling of mobile ad hoc networks. Without taking advantage of mobility, they exploited multiuser detection among other signal processing techniques to enhance user capacity.

2.4   Infrastructure Matters: Capacity of Hybrid Wireless Networks

Unlike pure ad hoc networks of homogeneous nodes operating in the same manner, hybrid wireless networks consist of at least two types of nodes functioning differently. After [22], significant efforts have been made to investigate capacity and delay scaling considering node heterogeneity, i.e., for hybrid networks, including wireless ad hoc networks with infrastructure aiding nodes, ad hoc networks with wireless helping nodes, multihop acess networks, and cognitive radio networks, among others.

Table 2.2 Scaling regimes shown in [101]

Regime	Number of infrastructure nodes	Per-node throughput capacity
(i)	$M \lesssim \sqrt{N/\log N}$	$\Theta(1/\sqrt{N \log N})$
(ii)	$\sqrt{N/\log N} \lesssim M \lesssim N/\log N$	$\Theta(M/N)$
(iii)	$M \gtrsim N/\log N$	$\Theta(1/\log N)$

2.4.1 Ad Hoc Networks with Supportive Infrastructure

It has been shown that adding wired infrastructure nodes, such as base stations, to ad hoc networks can render significant benefits in terms of both throughput capacity and delay. In the context of related investigations, the fixed infrastructure supports the underlying ad hoc networks by relaying their packets, rather than access points to the Internet. The advantage of infrastructure nodes is to overcome geographic limitations since the packet can be relayed over a long distance through high-bandwidth wired links, as a complement of local ad hoc delivery.

Liu et al. [98] initiated the study on capacity scaling of hybrid wireless networks. By placing N stationary nodes and M base stations in the network, they found that the throughput capacity increases linearly with M if $M = \omega(\sqrt{N})$, otherwise the improvement is negligible. Different from the hexagonal cell structure of base station in [98], access points in [99] are randomly distributed in the network and the results show that it is possible to achieve a throughput of $\Theta(\frac{1}{\log N})$ under the condition that the number of ad hoc nodes associated with each access point is upper bounded. In [100], Toumpis derived capacity bounds of hybrid wireless networks assuming randomly located access points and a general fading channel model and reported very similar results to those in [98]. In [101], Zemlianov et al. provided upper bounds of per-node throughput capacity for the network of randomly distributed ad hoc nodes and base stations placed in any deterministic fashion. By allowing power control of base stations, they determined three scaling regimes as shown in Table 2.2. It can be seen that there is no need to deploy any infrastructure for regime (i), since the throughput is achievable by only leveraging ad hoc communications; and for regime (iii), adding more infrastructure nodes does not make any improvement in throughput, at least in the order sense.

By noting that previous studies usually consider a two-dimensional square or disk network area, Liu et al. [102] investigated the impact of network geometry on capacity scaling by exploring one-dimensional networks and two-dimensional strip networks with regularly placed base stations. The main implications of theirs results (shown in Table 2.3) are: (i) for the one-dimensional network, even a small number of supportive base stations can significantly increase the per-node throughput capacity; and (ii) for a two-dimensional strip network, depending on the width of the strip, the behavior of capacity scaling is the same as that of either the one-dimensional network or the two-dimensional square network. The upper bound of average packet delay for each type of network was also derived, as shown in Table 2.3. Impacts of both network topology and traffic pattern were considered

Table 2.3 Impact of network geometry [102]

Network geometry	Number of base stations	Throughput capacity	Average delay
1-D network &	$M \log M = O(N)$	$\Omega(M/N)$	$O(N/M \log N)$
2-D strip with strip width of $o(\log N)$	$M \log M = \omega(N)$	$\Omega(1/\log M)$	$O(N/M \log N)$
2-D square &	$M = O(\sqrt{N})$	$\Omega(1/\sqrt{N})$	$O(\sqrt{N})$
2-D strip with strip width of $\Omega(\log N)$	$M = \omega(\sqrt{N})$	$\Omega(\min\{M/N, 1/\log M\})$	$O(\sqrt{N/M \log N})$

Table 2.4 Scaling regimes shown in [104]

Regime	Number of base stations	Throughput capacity	Average delay
(i)	$M = O(N/\log N)$	$\Omega(\sqrt{\frac{M}{N \log N}})$	$\Omega(\sqrt{\frac{N}{M \log N}})$
(ii)	$M = \Omega(N/\log N)$	$\Omega(M/N)$	$O(1)$

in [103]. Traffic patterns differ from each other in number of destination nodes in the network. The capacity scaling is determined by the number of base stations, the shape of network area, and the traffic pattern. Moreover, the impact of base station placement, i.e., regular or random placement, was also considered in [103].

An important implication of results shown in [98–103] is that capacity gain will be insignificant if the number of infrastructure nodes placed in a square or disk network area grows asymptotically slower than certain threshold. By pointing out that such "threshold" comes from the underutilization of the capability of base stations, Shila et al. [104] provided a better capacity and delay scaling, as shown in Table 2.4. The basic strategy they adopted is to deliver a packet to the nearest base station through multiple hops, in contrast to the one-hop transmission from the node to the associated base station assumed in previous studies, which results in a sublinear capacity scaling with the number of base stations.

Li et al. [105] revisited capacity and delay scaling in hybrid wireless networks by exploiting an *L-maximum-hop routing strategy*. Specifically, if the destination can be reached within L hops, packets from the source are delivered without reliance on any infrastructure node. More importantly, it was shown that without degrading throughput, network delay can be improved substantially, however, at the expense of built infrastructure. It is possible to achieve both constant throughput and delay in such type of networks. By using the L-maximum-hop routing strategy as well,

Zhang et al. [106] studied the throughput capacity for a network of N randomly distributed nodes, each of which is equipped with a directional antenna, and M regularly placed base stations. By analyzing the relationship between L, M and directional antenna beamwidth θ, they showed a "threshold" result on impacts of directional antenna, i.e., throughput gain can be achieved by implementing directional antenna only when the number of base stations grows slower than certain threshold. Multiantenna systems were also considered. In [107], Shin et al. investigated the capacity scaling in the network with supportive base stations, at each of which the number of antennas scales at arbitrary rates relative to N. It is beneficial to exploit the spatial dimension of infrastructure by deploying multiple antennas, which enable simultaneous uplinks, at each base station. The throughput capacity of mobile hybrid networks was reported in [108], in which the mobility model considered is similar to that in [95].

2.4.2 Ad Hoc Networks with Wireless Aiding Nodes

Deploying wired infrastructure to support ad hoc networks may incur a large amount of investment. Cost is always an important concern of building real-world communication networks. Moreover, under some emergency (e.g., earthquake) or extreme (e.g., underwater) circumstances, infrastructure is typically unavailable. Therefore, a potential substitute is to deploy a set of aiding nodes which are wirelessly connected and more powerful than normal nodes, as shown in Fig. 2.5. A natural question arises in the context: how much capacity gain can be achieved? To answer this question, Li et al. [109] studied the throughput capacity of ad hoc networks with the deployment of wireless helping nodes. Other specific network features considered in [109] are rectangular network area, both regular and random placement of helping nodes, and asymmetric traffic in which the number of destination nodes can scale at a lower rate than $\Theta(N)$, all of which have large impacts on throughput capacity. The main result of [109] illustrates that it is possible to achieve higher per-node throughput than that of pure ad hoc networks when the allocated bandwidth of helping nodes scales at a much higher rate than $\Theta(1)$. In [110], Zhou et al. provided another promising solution of wireless mesh structures. In such hierarchical wireless mesh network, mesh clients (normal nodes) are uniformly distributed, and mesh routers (aiding nodes) constitute a wireless mesh backbone, some of which can function as infrastructure gateways. Asymptotic throughput was derived and represented by the number of mesh clients, the number of mesh routers and the number of mesh gateways. Relying on only a small number of mesh gateways, it was shown that such mesh network can achieve the same throughput capacity as that of a hybrid infrastructure-based network, however, with a much lower cost.

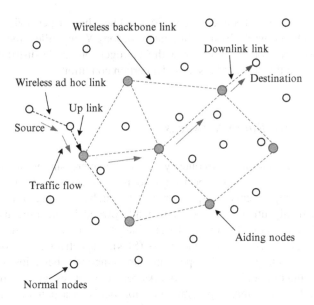

Fig. 2.5 Ad hoc network supported by wirelessly connected aiding nodes

2.4.3   Multihop Access Networks

Unlike ad hoc networks with supportive infrastructure nodes which do not generate or consume any data traffic, multihop access networks consist of infrastructure gateways bring/routing data traffic from/to the outside, such as Internet. Moreover, ad hoc transmissions between normal nodes are enabled and expected to enhance performance of such access networks, including capacity, coverage, connectivity, among others. To justify the benefit of augmenting access networks with multihop wireless links, Law et al. [111] investigated the downlink capacity of multihop cellular networks with regular placement of normal nodes and base stations. Due to poor spatial reuse, it was shown that one-dimensional multihop cellular networks yield almost no capacity gain compared to pure cellular networks. However, it is possible to significantly improve capacity of hexagonal hybrid network by exploiting multihop wireless links. By analyzing mathematically, they also found that capacity scaling in such type of networks mainly depends on the coverage of the base station, the transmission range of ad hoc links, and bandwidth allocation between different types of links. As a follow-up effort, Li et al. in [112] investigated capacity scaling for multihop cellular networks of randomly placed base stations and normal nodes distributed following a general inhomogeneous poisson process (IPP). In addition, throughput capacity was analyzed under different fairness constraints: (i) throughput-fairness, making throughput equal over all the nodes; and (ii) bandwidth-fairness, which guarantees that each node has equal allocated bandwidth. A "$\log_2 N$" result was shown in [112], i.e., multihop cellular networks with regular placement of nodes and base stations achieve higher per-node throughput than pure cellular networks by a scaling factor of $\log_2 N$, regardless the underlying fairness constraint. For the network with heterogeneous node distribution, it is possible to obtain the "$\log_2 N$" result under certain conditions.

2.4.4 Cognitive Radio Networks

Nowadays, the demand on the frequency spectrum is increasingly difficult to meet due to scarce and underutilized spectrum resources. Cognitive radio is a paradigm created in an attempt to enhance spectrum utilization, by enabling unlicensed users to opportunistically utilize the spectrum bands owned by licensed users [113]. In cognitive radio networks, licensed users and unlicensed users are referred to as primary users (PUs) and secondary users (SUs), respectively. With overlapping primary and secondary networks operating simultaneously, behaviors of capacity and delay scaling of cognitive radio networks need to be investigated carefully.

By only allowing single-hop communication between a pair of SUs, Vu and Tarok [114] showed that the aggregate throughput of SUs can scale linearly with the number of SUs in the presence of a single or multiple pairs of primary transmitter (TX) and receiver (RX). In [115], Jeon et al. considered an ad hoc primary network of N randomly distributed PUs overlapped with an ad hoc secondary network of M randomly distributed SUs. Assuming M is much larger than N, they showed that an aggregate throughput of $\Theta(\sqrt{N})$ is achievable for the primary network, and in the meantime, the aggregate throughput of the secondary network is $\Theta(M^{\frac{1}{2}-\delta})$, for any arbitrarily small fraction of outage δ. The main implication of their result is that both two networks have almost the same capacity scaling as if each were a single network, given that one is much denser than the other. Another assumption made in [115] is that SUs know the locations of primary RXs. However, they are typically unavailable in practical scenarios. Instead, Yin et al. [116] studied capacity scaling of cognitive radio networks on the assumption that the locations of primary TXs are available to SUs and obtained very similar results to those in [115]. Huang and Wang [117] considered a more general model of cognitive radio networks, where the primary network can be different types, including classic static network, network with random walk mobility, and hybrid network, among others. Within this scope, they showed that the secondary network can attain the same asymptotic capacity and delay as standalone networks. The literature [118] is different from previous works in twofold. First, SUs are mobile and follow a specific heterogeneous speed-restricted mobility model. Second, cooperative communications are enabled so that SUs are allowed to relay packets for PUs. By exploiting the mobility heterogeneity of SUs, it was shown that almost constant capacity and delay scalings (except for poly-logarithmic factors) are possible in such kind of cognitive radio networks.

2.5 Summary

This chapter has surveyed the existing literature for scaling laws of throughput capacity for both ad hoc wireless networks and wireless networks with communication infrastructure. It has also presented a comprehensive overview of capacity-delay tradeoffs under a variety of mobility models. Extensive comparisons of existing results have been done to reach a better understanding.

Chapter 3
Unicast Capacity of Vehicular Networks with Socialized Mobility

In this chapter, we investigate the throughput capacity of social-proximity vehicular networks. The considered network consists of N vehicles moving and communicating on a scalable grid-like street layout following the social-proximity model: each vehicle has a restricted mobility region around a specific social spot, and transmits via a unicast flow to a destination vehicle which is associated with the same social spot. Furthermore, the spatial distribution of the vehicle decays following a power-law distribution from the central social spot towards the border of the mobility region. With vehicles communicating using a variant of the two-hop relay scheme, the asymptotic bounds of throughput capacity are derived in terms of the number of social spots, the size of the mobility region, and the decay factor of the power-law distribution. By identifying these key impact factors of performance mathematically, we find three possible regimes for the throughput capacity and show that inherent mobility patterns of vehicles have considerable impact on network performance.

3.1 Introduction

We consider the following challenging features of VANETs in this chapter. *Firstly*, in VANETs, vehicles have **map-restricted** and **localized** mobility with specific social features. Notably, for most of the time, a vehicle only moves within a bounded region related to the social life of the driver. For example, a vehicle often moves within a small area daily which is close to the driver's home, the work place, or the city center. This mobility feature has also been reported in [119] based on the analysis of the real-world mobility trace of taxis in the city of Warsaw, Poland. It is observed that the mobility of taxis is typically around certain social spots. *Secondly*, VANETs show high **spatial variations** of vehicle density [3]. The analysis of the Warsaw trace data in [119] also reveals that the density of vehicles within the proximity area of social spots is much higher than on average

N. Lu and X. Shen, *Capacity Analysis of Vehicular Communication Networks*, 27
SpringerBriefs in Electrical and Computer Engineering,
DOI 10.1007/978-1-4614-8397-7_3, © The Author(s) 2014

and follows the empirical heavy-tailed distribution. *Thirdly*, VANETs serve many **proximity-related** applications, such as safety message dissemination and localized social content sharing, and thereby it is neither practical nor necessary to maintain a long-lasting unicast communication flow among vehicles over a long-distance. The in-depth investigations on throughput capacity are very limited, especially with the specific features aforementioned. However, it is desirable to know the fundamental capability of such social-proximity VANETs, which motivates our work (Refer to Sect. 2.3.3 for related works).

In this chapter, we investigate the throughput capacity of the social-proximity urban VANET. In specific, we model the urban area as a scalable grid with equal-length road segments and a set of social spots. Considering the localized and social features of vehicle's mobility, we apply a restricted mobility region to each vehicle centering at a fixed social spot with the spatial stationary distribution of the vehicle following a power-law decay from the social spot to the border of the mobility region. Based on this network model, we consider the social-proximity applications such that the data traffic is delivered through unicast flows; and for each unicast flow, its source and destination vehicles belong to the same social spot. Applying a variant of the two-hop relay scheme [84], we derive the bounds of throughput capacity and show how the asymptotic results depend on the inherent mobility pattern of the network which is characterized by the number of social spots, size of the mobility region, and the decay factor of the spatial distribution. The main contributions are three-fold:

• Our work represents the first theoretical study on the social-proximity VANET. As vehicular communications are intensively affected by the social behaviors of drivers, we argue that to accurately model the social features of vehicle mobility is crucial for the study of vehicular communications.
• We provide a generic modeling framework to derive the asymptotic throughput capacity of the social-proximity VANET, which can be used to predict network performance and provide guidance on network design and deployment.

The remainder of this chapter is organized as follows: In Sect. 3.2, we introduce the system models. Section 3.4 summarizes the main results of this work. We analyze the asymptotic throughput capacity with the proposed two-hop relay scheme in Sect. 3.3. Section 3.5 gives a brief summary.

3.2 System Model

3.2.1 Street Pattern

The geographic area of the network is modeled as a grid-like street layout, which consists of a set of M vertical roads intersected with a set of M horizontal roads, as shown in Fig. 3.1. Each line segment of equal length represents a road segment

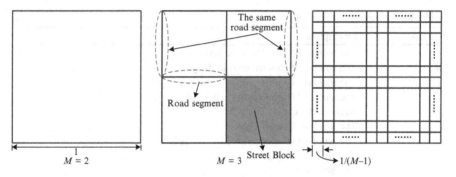

Fig. 3.1 A grid-like street layout

with bi-directional vehicle traffic. The grid street pattern is very common in urban metropolitan areas, such as Houston and Portland [120]. In the model, M is used to characterize the scale of the city grid. For example, M is roughly 100 for the downtown area of Toronto [121]. To eliminate the border effects, the city grid is considered as a torus of unit area, which is a common practice to avoid tedious technicalities [95].

Let C denote the number of street blocks in the grid. The total number of road segments (the road section between any two neighboring intersections) is therefore $G = 2C = 2(M-1)^2$. We define the *network density* $\psi = \frac{N}{G} = \frac{N}{2(M-1)^2}$, where N is the total number of vehicles moving on the roads. Since N tends to infinity in the asymptotic study, the city size, determined by M, should be scalable as well. Let $\Theta(1) \leq \psi \leq o(N)$ to avoid two extreme cases which are not practical in real-world scenarios: (1) when $\psi = o(1)$, the city size increases faster than the population of vehicles; and (2) when $\psi = \Theta(N)$, the city size is fixed such that the network density becomes extremely high when more and more vehicles appear in the city. Note that ψ represents the average vehicle density on each road segment. However, as each vehicle moves following the mobility model with social features, the spatial distribution of vehicles is inhomogeneous, as examples shown in Fig. 3.3b, c. It can be seen that a network with a very large M and a relatively large ψ can represent metropolitan areas like New York City; whereas for a small town, M and ψ are relatively small. Therefore, from a macroscopic view, the grid street pattern with different values of M and ψ can model urban scenarios of different scales. A summary of the mathematical notations used in this chapter is given in Table 3.1.

3.2.2 Socialized Mobility Model

Markovian Mobility Pattern: We consider a time-slotted communication system, where time is slotted with equal duration. The road segments are indexed from 1 to G and vehicle nodes are indexed from 1 to N. Vehicles move independently from

Table 3.1 The notations for Chap. 3

Symbol	Description
N	The number of vehicles in the network
M	The number of parallel roads in the grid
G	The number of road segments in the grid
C	The number of street blocks in the grid
ψ	Network density
V	The number of social spots
$\mathbf{S}$	The set of social spots, $\mathbf{S} = \{S_1, \ldots, S_V\}$
H_k	Vehicle k's social spot, $H_k \in \mathbf{S}$
ν	Exponent of V
κ	Exponent of $\mathscr{A}$
$\mathscr{A}$	The outermost tier of vehicle's mobility region
π'_α	The steady-state location probability of each vehicle on a segment of $Tier(\alpha)$
γ	The decay factor
r	The communication radius of the vehicle
Δ	The guard factor
Φ	The family of social-proximity VANETs
$\lambda(\Phi)$	Per-vehicle throughput
$\tilde{\lambda}(\Phi)$	Average per-vehicle throughput
p_{ac}	The probability of a randomly selected road segment being active at a time slot
d_i^N	The vehicle density of road segment i
$\underline{d}_i^N$	The lower bound of d_i^N
$\overline{d}_i^N$	The upper bound of d_i^N
F_i	The rectangular area of $2\mathscr{A}(2\mathscr{A} - 1)$ street blocks centered at road segment i
F_i^s	The number of social spots in F_i
S_j^v	The number of S-D pairs associated with social spot S_j
$\mathscr{N}_i$	The number of vehicles on road segment i during a time slot
$\mathscr{N}_i^{SD}$	The number of S-D pairs on road segment i during a time slot
$P(\Phi)$	The average number of road segments where there are at least two vehicles during a time slot
$Q(\Phi)$	The average number of road segments where there is at least one S-D pair during a time slot

each other in the city. The mobility of a vehicle k follows a discrete time Markovian process, denoted by $\mathscr{C}_k, k \in \{1, 2, \ldots, N\}$, which is uniquely represented by a one-dimensional G-state ergodic Markov Chain. $\mathscr{C}_k(t) = i$ if vehicle k appears on road segment i, $i \in \{1, 2, \ldots, G\}$, at time slot t, $t \in \{1, 2, \ldots, T\}$. Let P_k^{ij} denote the transition probability that vehicle k moves from road segment i to the next road segment j, $j \in \{1, 2, \ldots, G\}$. Let $\mathbf{P}_k = \{P_k^{ij}\}_{G \times G}$ denote the transition probability matrix of $\mathscr{C}_k$; the element P_k^{ij} in $\mathbf{P}_k$ is non-zero only if j is a neighboring road segment of i. The steady-state location distribution of vehicle k is $\boldsymbol{\pi}_k = \{\pi_k(i)\}_{1 \times G}$, where $\pi_k(i)$ denotes the long-term proportion of time that vehicle k stays on road segment i. In our analysis, we focus on the steady-state location distribution of the vehicles, since the capacity region only depends on how the node location distributes in the steady state [122], and the Markovian mobility model converges to its steady-state location distribution at an exponential rate [123].

Restricted Mobility Region with Social Spot: The mobility region of each vehicle is restricted and associated with a fixed social spot. Geographically, the social spot is the center of a certain street block, as shown in Fig. 3.2. Let V denote the number of social spots in the grid. We assume that all the social spots in the grid are uniformly distributed and thereby do not consider the inhomogeneous distribution of social spots in this study. Indexing all the street blocks from 1 to C, we denote by $\mathbf{S} = \{S_1, S_2, \ldots, S_V\} \subseteq \{1, 2, \ldots, C\}$ the set of social spots. Since we are interested in the capacity order, let $V = |\mathbf{S}| = \lceil C^\nu \rceil$, where $\nu \in (0, 1]$. Notably, $V = \Theta((N/\psi)^\nu)$, which is represented by a power function of N. When $\nu = 1$, all the street blocks in the network contain a social spot. In addition, we consider all the intermediate cases of ν between 0 and 1 in this research.[1] Each vehicle uniformly and independently selects one social spot out of all the social spots. Let $\mathbf{H}_N = (H_1, H_2, \ldots, H_N)$ denote the vector which collects the locations of all the vehicles' social spots, with each element $H_k \in \mathbf{S}$, denoting the index of the street block where vehicle k's social spot is located. The set $\mathbf{S}$ is fixed once the network is defined.

The mobility region of each vehicle consists of multiple tiers co-centered at its social spot, as shown in Fig. 3.2. $Tier(1)$ of the mobility region is collocated with the social spot and contains four road segments. The adjacent street blocks surrounding $Tier(1)$ form $Tier(2)$, and so on. We denote by $Tier(\mathscr{A})$ the outermost tier of the mobility region, where $\mathscr{A} = \Theta(M^\kappa) = \Theta((N/\psi)^{\frac{\kappa}{2}}) \leq \lfloor \frac{M}{2} \rfloor$, $\kappa \in [0, 1)$. When $\kappa = 0$, the size of the mobility region is fixed and does not scale with the city grid. It can be easily derived that $Tier(\alpha)$, $\alpha \in \{1, 2, \ldots, \mathscr{A}\}$, contains $16\alpha - 12$ road segments. Thus, the mobility of each vehicle is constrained in $\mathscr{A}$ tiers with a total number of $\sum_{\alpha=1}^{\mathscr{A}} 16\alpha - 12 = 4\mathscr{A}(2\mathscr{A} - 1)$ road segments, and further the mobility region covers an area of $(\mathscr{A} + 1)^2/C = \Theta((N/\psi)^{\kappa-1})$. For a randomly selected $Tier(\alpha)$, a vehicle has equal steady-state probability to appear on each road segment. Let π'_α denote the steady-state location probability of each vehicle on one of the road segments of $Tier(\alpha)$. From $Tier(1)$ to $Tier(\mathscr{A})$, the steady-state location probability of vehicles is modeled to exponentially decay as a power-law function with exponent $\gamma > 0$. Therefore, we have $\pi'_\alpha = \alpha^{-\gamma} \pi'_1$ which indicates that a vehicle is more likely to stay in the area near its social spot. The same model has been used in [95] and its accuracy is validated in [119] through real-world measurements. As the summation of steady-state probability on road segments equals to 1, i.e.,

---

[1]We do not consider the extreme case in which $\nu = 0$. When $\nu = 0$, there is only one social spot in the network.

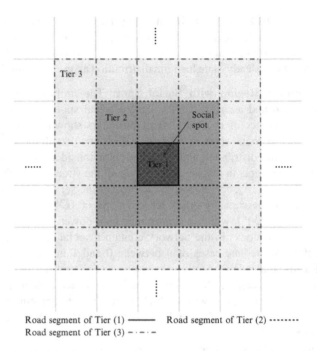

Road segment of Tier (1) ———— Road segment of Tier (2) ·········
Road segment of Tier (3) — · — · —

Fig. 3.2 Restricted and socialized mobility with different tiers centered at a social spot for a given vehicle

$$\sum_{\alpha=1}^{\mathscr{A}}(16\alpha - 12)\pi'_\alpha = \sum_{\alpha=1}^{\mathscr{A}}(16\alpha - 12)\alpha^{-\gamma}\pi'_1 = 1,$$

we have,

$$\pi'_1 = \frac{1}{\sum_{\alpha=1}^{\mathscr{A}}(16\alpha - 12)\alpha^{-\gamma}}. \tag{3.1}$$

Lemma 3.1. *Given that $\kappa > 0$, as $N \to \infty$, $\pi'_1 = \Theta((N/\psi)^{-\kappa(1-\frac{1}{2}\gamma)})$, for $0 < \gamma < 2$; $\pi'_1 = \Theta(\frac{1}{\log(N/\psi)})$, for $\gamma = 2$; π'_1 converges to a constant value, for any $\gamma > 2$.*

This lemma can be proved by applying results of partial sums of p-series [124]. Note that when $\kappa = 0$, π'_1 is constant for all γ. Under the socialized mobility model, the network presents inhomogeneous vehicle densities. Figure 3.3 illustrates the vehicle density when vehicles are uniformly distributed and follow the socialized mobility model, respectively.

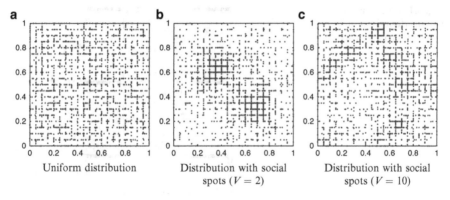

Fig. 3.3 Examples of homogeneous (**a**) and inhomogeneous (**b**) and (**c**) distributions of vehicles in the network, in the case of $N = 2,000$, $M = 21$, $\mathscr{A} = 10$ and $\gamma = 2$

3.2.3 Traffic Model

We consider that there exist N unicast flows concurrently in the network. Each vehicle is the source of one unicast flow and the destination of another unicast flow. We consider the case in which the source and destination vehicles of each unicast flow have the same social spot. This is motivated by the dominant proximity applications in vehicular communications. By doing so, the source and destination vehicles of each unicast flow are spatially close to each other. Without loss of generality, N is considered to be even. We sort the index of vehicles such that vehicle k communicates with vehicle $k + 1$, $k \in \{1, 3, 5, \ldots, N - 1\}$, and each communication pair independently and uniformly chooses a social spot from **S**. The packet arrives in each unicast flow at an average rate η.

3.2.4 Communication Model

Since the communication range of a vehicle is geographically limited in practice, the communication radius should scale with M. Let $r = \frac{1}{M-1}$ denote the communication radius of each vehicle which can always cover the entire road segment, as shown in Fig. 3.4. Without loss of generality, a pair of vehicles can communicate only when they are on the same road segment at the same time slot, and the transmission spans the whole time slot. Although the communication model has been simplified, such simplification does not affect the order of throughput capacity and average packet delay derived in this chapter. The success or failure of a transmission is determined by the protocol model defined in [22] as follows. The transmission from vehicle i to vehicle j can be successful during time slot t if and only if the following condition holds: $d_{kj}(t) \geq (1 + \Delta)r$, for every other vehicle k transmitting simultaneously, where $d_{kj}(t)$ denotes the Euclidean distance between vehicle k and j at time slot t, and $\Delta > 0$ is the guard factor.

Fig. 3.4 An example of non-interfering transmission group of road segments

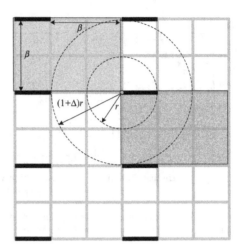

3.2.5 Definitions

We denote by $\boldsymbol{\Phi}(N, \psi, \gamma, \mathscr{A}(\kappa), \mathbf{S}(\nu), \mathbf{H}_N)$ the family of social-proximity VANETs. Let $L_k(T)$ be the number of packets received by the destination of flow k, $k \in \{1, 2, \ldots, N\}$, up to time T. An asymptotic per-vehicle throughput $\lambda(\boldsymbol{\Phi})$ of $\boldsymbol{\Phi}$ is said feasible if there exist a scheduling policy and an N_0 such that for any $N > N_0$, we have,

$$\lim_{T \to \infty} \mathbf{Pr}\left(\frac{L_k(T)}{T} \geq \lambda(\boldsymbol{\Phi}), \forall k \right) = 1. \tag{3.2}$$

Furthermore, an average per-vehicle throughput $\tilde{\lambda}(\boldsymbol{\Phi})$ of $\boldsymbol{\Phi}$ is said feasible if there exist a scheduling policy and an N_0, such that for any $N > N_0$, the following holds

$$\lim_{T \to \infty} \mathbf{Pr}\left(\frac{\sum_{k=1}^{N} L_k(T)}{NT} \geq \tilde{\lambda}(\boldsymbol{\Phi}) \right) = 1. \tag{3.3}$$

3.3 Asymptotic Capacity Analysis

In this section, a two-hop relay scheme is first proposed to delivery the packets from the source to the destination. We then derive the bounds of per-vehicle throughput capacity and average per-vehicle throughput, which are stated in Theorem 3.1 and 3.2, respectively.

3.3.1 Two-Hop Relay Scheme

All packets are transmitted by using a two-hop relay scheme $\mathscr{X}$: a packet is either transmitted directly from the source to the destination, or relayed through one intermediate vehicle from the source to the destination. The packet transmission consists of two phases:

> $\mathscr{X}$-I: Each road segment in the network becomes "active" in every $1/p_{ac}$ time slots.[2]
>
> $\mathscr{X}$-II: For each active road segment where there are at least two vehicles,

1. If there exists at least one source-destination (S-D) pair on the road segment, one pair is uniformly selected. If the source has a buffering packet for the destination, it transmits the packet and deletes it from the buffer after the transmission; otherwise, the source stays idle.
2. If there is no any S-D pair on the road segment, a vehicle, e.g., v_A, is uniformly selected out of all vehicles on this road segment to be the source or the destination equally likely, and in the meantime another vehicle, e.g., v_B, is independently and uniformly selected over the rest of vehicles to be the relay.

- If v_A is the source, a source-to-relay transmission from v_A to v_B is scheduled. If v_A has a buffering packet to transmit, v_A transmits the packet to v_B and deletes the packet from the buffer; otherwise, v_A remains idle.
- If v_A is the destination, a relay-to-destination transmission from v_B to v_A is scheduled. If v_B has a buffering packet destined for v_A, v_B transmits the packet to v_A and deletes the packet from the buffer; otherwise, v_B remains idle.

We calculate the value of p_{ac} in the following, which is the probability of a randomly selected road segment being active at a time slot. As shown in Fig. 3.4, we partition the network into equal-size sub-areas. Each sub-area consists of $\beta(\beta + 1)$ street blocks where β is an integer number. The road segments highlighted in each sub-area in Fig. 3.4 constitute a non-interfering transmission group, such that simultaneous transmissions within one non-interfering group do not interfere with each other. Totally, there are $2\beta(\beta + 1)$ road segments within one sub-area, and collectively $2\beta(\beta + 1)$ non-interfering groups in the network. With non-interfering groups transmitting alternately, each non-interfering group becomes active every $1/p_{ac} = 2\beta(\beta + 1)$ time slots. This indicates that the vehicles on one specific road segment obtain a transmission opportunity with probability p_{ac} at a randomly selected time slot. With the grid scale of M, the minimum distance between any two neighboring road segments of a non-interfering group is $\frac{\beta}{M-1}$. By using the protocol model, we have

---

[2]A road segment is active when vehicles on the road segment can transmit successfully without any interference of transmissions from other road segments. The value of p_{ac} is discussed later in the section.

$$\beta/(M-1) \geq (1+\Delta)r.$$

With $r = 1/(M-1)$, we have $\beta \geq 1 + \Delta$. We set $\beta = \lceil 1 + \Delta \rceil$. By substituting it into $1/[2\beta(\beta+1)]$, we have

$$p_{ac} = 1/(2\lceil 1+\Delta \rceil \lceil 2+\Delta \rceil).$$

3.3.2   Bounds of Per-Vehicle Throughput Capacity

In the following, we derive the bounds of the per-vehicle throughput capacity by using the two-hop relay scheme $\mathscr{X}$, which are formally stated in Theorem 3.1. We first obtain an important result of vehicle density of a generic road segment (Lemma 3.3) by applying Chernoff bounds (Lemma 3.2) and the Vapnik-Chervonenkis Theorem which gives the uniform convergence in the weak law of large numbers.

To characterize the vehicle spatial inhomogeneities of the network, inspired by [93], we define the vehicle density (*vehicles/road segment*) of a generic road segment i by

$$d_i^N = \sum_{k=1}^{N} E[\mathbb{I}_{\mathscr{C}_k(t)=i} | \mathbf{H}_N]. \tag{3.4}$$

where $\mathbb{I}_{\mathscr{C}_k(t)=i}$ is the indicator variable that takes value 1 if $\mathscr{C}_k(t) = i$ and 0 otherwise.

Lemma 3.2. *(Chernoff bounds [125]) Let X be a sum of n independent random variables $\{X_i\}$, with $X_i \in \{0,1\}$ for all $i \leq n$. Write $\mu' = E[X] = E[X_1] + \cdots + E[X_n]$. Then for any $0 < \varepsilon \leq 1$,*

$$\mathbf{Pr}(X > (1+\varepsilon)\mu') \leq e^{-\frac{\varepsilon^2}{2+\varepsilon}\mu'}, \quad and$$

$$\mathbf{Pr}(X < (1-\varepsilon)\mu') \leq e^{-\frac{\varepsilon^2}{2}\mu'}.$$

Lemma 3.2 is a well-known result and used to prove the following important lemma which presents a bound of d_i^N.

Lemma 3.3. *The following bounds of the vehicle density d_i^N hold w.h.p.,[3] $\forall i \in \{1, 2, \ldots, G\}$:*

(i) When $\kappa + \nu > 1$,

---

[3] As $N \to \infty$, the probability of the event approaches 1.

Fig. 3.5 An example of one given road segment contained by different vehicles' mobility regions

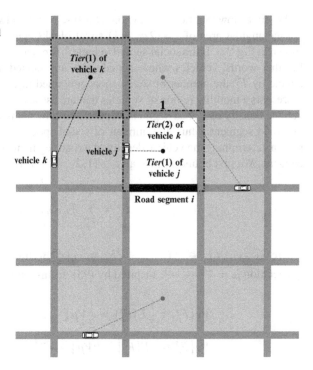

$$
d_i^N = \begin{cases} \Omega(\psi), \ O(\psi(N/\psi)^{\frac{1}{2}\kappa\gamma}) & 0 < \gamma < 2 \\[2mm] \Omega(\frac{\psi}{\log(N/\psi)}), \ O(\frac{\psi(N/\psi)^{\kappa}}{\log(N/\psi)}) & \gamma = 2 \\[2mm] \Omega(\frac{\psi}{(N/\psi)^{\kappa(\frac{1}{2}\gamma-1)}}), \ O(\psi(N/\psi)^{\kappa}) & \gamma > 2. \end{cases}
$$

(ii) When $\kappa + v \le 1$,

    (a) $\kappa = 0$, $d_i^N = O(\psi(N/\psi)^{1-v} \log(N/\psi))$;

    (b) $\kappa \ne 0$,

$$
d_i^N = \begin{cases} O\left(\frac{\psi \log(N/\psi)}{(N/\psi)^{v+\kappa(1-\frac{1}{2}\gamma)-1}}\right) & 0 < \gamma < 2 \\[2mm] O(\psi(N/\psi)^{1-v}) & \gamma = 2 \\[2mm] O(\psi(N/\psi)^{1-v} \log(N/\psi)) & \gamma > 2. \end{cases}
$$

Proof. The proof of Lemma 3.3 consists of three parts in terms of different values of v and κ.

(i) $0 < v < 1$ and $\kappa + v > 1$:

We first show that Lemma 3.3 holds for $0 < \nu < 1$ and $\kappa + \nu > 1$. Let F_i denote the rectangular area of $2\mathscr{A}(2\mathscr{A}-1)$ street blocks centered at road segment i, as shown in Fig. 3.5. If the social spot of vehicle k is not located in F_i, $E[\mathbb{I}_{\mathscr{C}_k(t)=i}] = 0$. In other words, vehicles whose social spots are located in F_i contribute d_i^N. We denote by F_i^s the number of social spots contained in F_i. Intuitively, a social spot represents a mobility region of vehicles that are associated with this social spot. The vehicle density of a road segment depends on how many mobility regions contain this road segment. Thus, the number of social spot in an area plays an important role in determining the vehicle density of a road segment, and further the throughput capacity. We first bound F_i^s, for all $i \in \{1, 2, \ldots, G\}$. By definition, we have,

$$F_i^s = \sum_{j=1}^{V} \mathbb{I}_{S_j \in F_i} \tag{3.5}$$

where $\mathbb{I}_{S_j \in F_i}$, $\forall j \in \{1, 2, \ldots, V\}$, are i.i.d Bernoulli random variables with expectation $\mu = \frac{2\mathscr{A}(2\mathscr{A}-1)}{C}$. Inspired by [93], from Lemma 3.2, we have

$$\mathbf{Pr}\{F_i^s < \frac{1}{2}E[F_i^s] = \frac{1}{2}V\mu\} < e^{-\frac{1}{8}V\mu}, \text{ and}$$

$$\mathbf{Pr}\{F_i^s > 2E[F_i^s] = 2V\mu\} < e^{-\frac{1}{3}V\mu} < e^{-\frac{1}{8}V\mu}.$$

Notably, $\mu = \Theta((N/\psi)^{\kappa-1})$. By applying the union bound, we have

$$\mathbf{Pr}\{\bigcap_i \{\frac{1}{2}E[F_i^s] < F_i^s < 2E[F_i^s]\}\}$$

$$= 1 - \mathbf{Pr}\{\bigcup_i \{F_i^s < \frac{1}{2}E[F_i^s] \cup F_i^s > 2E[F_i^s]\}\}$$

$$\geq 1 - \sum_{i=1}^{G} \mathbf{Pr}\{F_i^s < \frac{1}{2}E[F_i^s] \cup F_i^s > 2E[F_i^s]\}$$

$$\geq 1 - G2e^{-\frac{1}{8}V\mu}.$$

For $\kappa + \nu > 1$, $G2e^{-\frac{1}{8}V\mu} \to 0$, as $N \to \infty$. Hence, w.h.p., we get

$$\frac{1}{2}E[F_i^s] < F_i^s < 2E[F_i^s], \ \forall i \in \{1, 2, \ldots, G\}. \tag{3.6}$$

Let $S_j^\nu = \sum_{k=1}^{N-1} \mathbb{I}_{H_k=S_j}$ denote the number of S-D pairs associated with social spot S_j, where $\mathbb{I}_{H_k=S_j}$, $\forall k \in \{1, 3, \ldots, N-1\}$, are i.i.d Bernoulli random variables with expectation $1/V$. From Lemma 3.2, for $0 < \varepsilon \leq 1$, we have

$$\mathbf{Pr}\{\frac{(1-\varepsilon)N}{2V} < S_j^v < \frac{(1+\varepsilon)N}{2V}\} \geq 1 - 2e^{-\frac{\varepsilon^2 N}{2(2+\varepsilon)V}}.$$

By applying union bound in the same manner,

$$\mathbf{Pr}\{\bigcap_j \{\frac{(1-\varepsilon)N}{2V} < S_j^v < \frac{(1+\varepsilon)N}{2V}\}\}$$

$$= 1 - \mathbf{Pr}\{\bigcup_j \{S_j^v < \frac{(1-\varepsilon)N}{2V} \cup S_j^v > \frac{(1+\varepsilon)N}{2V}\}\}$$

$$\geq 1 - \sum_{j=1}^{V} \mathbf{Pr}\{S_j^v < \frac{(1-\varepsilon)N}{2V} \cup S_j^v > \frac{(1+\varepsilon)N}{2V}\}$$

$$\geq 1 - V2e^{-\frac{\varepsilon^2 N}{2(2+\varepsilon)V}}.$$

Since $V = \Theta((N/\psi)^v)$ and $v \neq 1$, $V2e^{-\frac{\varepsilon^2 N}{(2+\varepsilon)V}} \to 0$, as $N \to \infty$. Hence, *w.h.p.*, we have

$$\frac{(1-\varepsilon)N}{2V} < S_j^v < \frac{(1+\varepsilon)N}{2V}, \quad \forall j \in \{1, 2, \ldots, V\}. \tag{3.7}$$

By definition (3.4) and from (3.6) and (3.7), *w.h.p.*, we obtain, $\forall i$,

$$\frac{1}{2}V\mu \cdot \frac{(1-\varepsilon)N}{V} \cdot \mathscr{A}^{-\gamma}\pi_1' = \underline{d}_i^N < d_i^N < \overline{d}_i^N$$

$$= 2V\mu \cdot \frac{(1+\varepsilon)N}{V} \cdot \pi_1'$$

denoting by $\underline{d}_i^N$ and $\overline{d}_i^N$ the lower bound and upper bound of d_i^N, respectively. Letting $\varepsilon \to \infty$, we have $\underline{d}_i^N = \Theta(\pi_1'\psi(N/\psi)^{\kappa(1-\frac{1}{2}\gamma)})$ and $\overline{d}_i^N = \Theta(\pi_1'\psi(N/\psi)^{\kappa})$. Then (i) follows according to Lemma 3.1.

(ii) $v = 1$:

To prove Lemma 3.3 for the case in which $v = 1$, we recall the Vapnik-Chervonenkis Theorem [126]. Some relevant definitions are first provided. A Range Space is a pair $(X, \mathscr{F})$, where X is a set and $\mathscr{F}$ is a family of subsets of X. For any $A \subseteq X$, we define $P_{\mathscr{F}}(A)$, the projection of $\mathscr{F}$ on A, as $\{F \cap A : F \in \mathscr{F}\}$. We say that A is *shattered* by $\mathscr{F}$ if $P_{\mathscr{F}}(A) = 2^A$, i.e., if the projection of $\mathscr{F}$ on A includes all possible subsets of A. The VC-dimension of $\mathscr{F}$, denoted by VC-d$(\mathscr{F})$ is the cardinality of the largest set A that $\mathscr{F}$ shatters. If arbitrarily large finite sets are shattered, the VC dimension of $\mathscr{F}$ is infinite.

The Vapnik-Chervonenkis Theorem: If $\mathscr{F}$ is a set of finite VC-dimension and $\{Y_j\}$ is a sequence of N i.i.d. random variables with common probability distribution P, then for every $\epsilon, \delta > 0$

$$\mathbf{Pr}\left\{ \sup_{F \in \mathscr{F}} \left| \frac{1}{N} \sum_{j=1}^{N} \mathbb{I}_{Y_j \in F} - P(F) \right| \leq \epsilon \right\} > 1 - \delta \tag{3.8}$$

if

$$N > \max\left\{ \frac{8\text{VC-d}(\mathscr{F})}{\epsilon} \log \frac{16e}{\epsilon}, \frac{4}{\epsilon} \log \frac{2}{\delta} \right\}. \tag{3.9}$$

We use the Vapnik-Chervonenkis Theorem to show that Lemma 3.3 holds for $v = 1$. Recall that F_i denotes the rectangular area of $2\mathscr{A}(2\mathscr{A} - 1)$ street blocks centered at road segment i. $\sum_{k=1}^{N-1} \mathbb{I}_{H_k \in F_i}, k \in \{1, 3, \dots, N - 1\}$, is the number of S-D pairs whose social spot falls into the region F_i. $\mathbf{Pr}(\mathbb{I}_{H_k \in F_i} = 1) = \frac{2\mathscr{A}(2\mathscr{A}-1)}{C} = \Theta((N/\psi)^{\kappa-1}), \forall k$. Let $\mathscr{F}$ be the class of all such F_i rectangular areas. It is easy to show that the VC-dimension of $\mathscr{F}$ is at most 4 [127]. Thus, $\forall F_i$,

$$\mathbf{Pr}\left\{ \sup_{F_i \in \mathscr{F}} \left| \frac{\sum_{k=1}^{N-1} \mathbb{I}_{H_k \in F_i}}{N/2} - \frac{2\mathscr{A}(2\mathscr{A} - 1)}{C} \right| \leq \epsilon \right\} > 1 - \delta.$$

The condition (3.9) holds when $\epsilon = \delta = \frac{\Delta_\epsilon \log (N/\psi)}{N/\psi}$, where $\Delta_\epsilon = \max\{8\text{VC-d}(\mathscr{F}), 16e\}$. Thus, the Vapnik-Chervonenkis Theorem states that

$$\mathbf{Pr}\left\{ \sup_{F_i \in \mathscr{F}} \left| \sum_{k=1}^{N-1} \mathbb{I}_{H_k \in F_i} - \Theta(\psi(\frac{N}{\psi})^\kappa) \right| \leq \Theta(\psi \log \frac{N}{\psi}) \right\}$$

$$> 1 - \frac{\Delta_\epsilon \log (N/\psi)}{N/\psi}.$$

We conclude that *w.h.p.*, for $\kappa = 0$,

$$\overline{d}_i^N = 2 \max\left\{ \sum_{k=1}^{N-1} \mathbb{I}_{H_k \in F_i} \right\} \pi_1' = \Theta(\psi \log(N/\psi)), \forall i;$$

for $0 < \kappa < 1$, $\underline{d}_i^N = \Theta(\pi_1' \psi (N/\psi)^{\kappa(1-\frac{1}{2}v)})$ and $\overline{d}_i^N = \Theta(\pi_1' \psi (N/\psi)^\kappa), \forall i$. Hence, (ii) follows according to Lemma 3.1.

(iii) $0 < v < 1$ and $\kappa + v \leq 1$:

We apply the Vapnik-Chervonenkis Theorem in the same manner. From (3.5), we have,

$$\mathbf{Pr}\left\{\sup_{F_i \in \mathscr{F}} \left| \frac{\sum_{j=1}^{V} \mathbb{I}_{S_j \in F_i}}{V} - \frac{2\mathscr{A}(2\mathscr{A}-1)}{C} \right| \le \epsilon \right\} > 1 - \delta.$$

The condition (3.9) holds when $\epsilon = \delta = \frac{\Delta_\epsilon \log(V)}{V} = \frac{\Delta_\epsilon \log(N/\psi)}{(N/\psi)^\nu}$. Thus,

$$\mathbf{Pr}\left\{ \sup_{F_i \in \mathscr{F}} \left| \sum_{j=1}^{V} \mathbb{I}_{S_j \in F_i} - \Theta((\frac{N}{\psi})^{\kappa + \nu - 1}) \right| \le \Delta_\epsilon \log(\frac{N}{\psi}) \right\}$$

$$> 1 - \frac{\Delta_\epsilon \log(N/\psi)}{(N/\psi)^\nu}. \qquad (3.10)$$

Since $\kappa + \nu \le 1$, we conclude that w.h.p., $F_i^s = O(\log(N/\psi))$, $\forall i$. From (3.7) and (3.10), the upper bound of vehicle density $\overline{d}_i^N = \Theta(\pi_1' \psi (N/\psi)^{1-\nu} \log(N/\psi))$, $\forall i$, w.h.p.. Then (iii) follows according to Lemma 3.1.

Theorem 3.1. *For the social-proximity grid-like VANETs, with the two-hop relay scheme* $\mathscr{X}$, *the per-vehicle throughput* $\lambda(\Phi)$ *cannot be better than* $\frac{1}{2\psi \lceil 1 + \Delta \rceil \lceil 2 + \Delta \rceil}$ *and w.h.p., we obtain*

(i) *When* $\kappa + \nu > 1$,

$$\lambda(\Phi) = \begin{cases} \Omega(\frac{1}{\psi(N/\psi)^{\frac{1}{2}\kappa\gamma}}) & 0 < \gamma < 2 \\ \Omega(\frac{1}{\psi(N/\psi)^\kappa \log(N/\psi)}) & \gamma = 2, \psi = \Theta(1) \\ \Omega(\frac{\log(N/\psi)}{\psi(N/\psi)^\kappa}) & \gamma = 2, \psi = \omega(1) \\ \Omega(\frac{1}{\psi(N/\psi)^\kappa}) & \gamma > 2. \end{cases}$$

(ii) *When* $\kappa + \nu \le 1$,

(a) $\kappa = 0$, $\lambda(\Phi) = \Omega(\frac{(N/\psi)^{\nu-1}}{\psi \log(N/\psi)})$;

(b) $\kappa \neq 0$,

$$\lambda(\Phi) = \begin{cases} \Omega(\frac{(N/\psi)^{\nu-\frac{1}{2}\kappa\gamma-1}}{\psi \log(N/\psi)}) & 0 < \gamma < 1 \\ \Omega(\frac{(N/\psi)^{\nu-\frac{1}{2}\kappa-1}}{\psi}) & \gamma = 1 \\ \Omega(\frac{(N/\psi)^{\nu-\kappa(1-\frac{1}{2}\gamma)-1}}{\psi \log(N/\psi)}) & 1 < \gamma < 2 \\ \Omega(\frac{(N/\psi)^{\nu-1}}{\psi \log^2(N/\psi)}) & \gamma = 2 \\ \Omega(\frac{(N/\psi)^{\nu-1}}{\psi \log(N/\psi)}) & \gamma > 2. \end{cases}$$

Proof. The proof consists of two parts. First, we apply Lemma 3.1 and 3.3 to derive the lower bound of the per-vehicle throughput. Following the two-hop relay scheme $\mathscr{X}$, the long-term throughput of flow k (denoting the source and destination of flow k by $\mathbb{S}$ and $\mathbb{D}$, respectively) is given by

$$
\lambda_k(\Phi) = \lim_{T \to \infty} L_k(T)/T
$$

$$
= \frac{1}{2} p_{ac} \sum_{i=1}^{G} \mathbf{Pr}(\mathcal{N}_i \geq 2, \mathcal{N}_i^{\mathrm{SD}} = 0|\mathscr{C}_{\mathbb{D}} = i)\pi_{\mathbb{D}}(i)\frac{1}{\mathcal{N}_i} \tag{3.11}
$$

$$
+ p_{ac} \sum_{i=1}^{G} \mathbf{Pr}(\mathscr{C}_{\mathbb{S}} = i|\mathscr{C}_{\mathbb{D}} = i)\pi_{\mathbb{D}}(i)\frac{1}{\mathcal{N}_i^{\mathrm{SD}}},
$$

where $\mathcal{N}_i$ and $\mathcal{N}_i^{\mathrm{SD}}$ denote the number of vehicles and the number of S-D pairs on road segment i in a time slot, respectively. Recall that $\pi_{\mathbb{D}}(i)$ is the steady-state probability that $\mathbb{D}$ stays on road segment i.

Let $\mathfrak{N}_i$ denote the number of S-$\mathbb{D}$ pairs whose mobility region contains road segment i. The probability of finding at least two vehicles and no any S-D pair on road segment i given that $\mathbb{D}$ is on road segment i is given by, *w.h.p.*, $\forall i$,

$$
\mathbf{Pr}(\mathcal{N}_i \geq 2, \mathcal{N}_i^{\mathrm{SD}} = 0|\mathscr{C}_{\mathbb{D}} = i)
$$

$$
\geq (1 - \pi_1')\left(1 - \left(1 - \frac{2\pi_1'}{\mathscr{A}^\gamma} + 2\left(\frac{\pi_1'}{\mathscr{A}^\gamma}\right)^2\right)^{\mathfrak{N}_i - 1}\right)
$$

$$
= (1 - \pi_1')\left(1 - \left(1 + \left(-\frac{\mathscr{A}^\gamma}{2\pi_1'}\right)^{-1}\right.\right.
$$

$$
\left.\left. + \frac{1}{2}\left(-\frac{\mathscr{A}^\gamma}{2\pi_1'}\right)^{-2}\right)^{-\frac{\mathscr{A}^\gamma}{2\pi_1'}(-2\mathfrak{N}_i\mathscr{A}^{-\gamma}\pi_1' + 2\mathscr{A}^{-\gamma}\pi_1')}\right)
$$

$$
\geq (1 - \pi_1')\left(1 - \left(\left(1 + \frac{1}{\mathscr{X}} + \frac{1}{2\mathscr{X}^2}\right)^{\mathscr{X}}\right)^{-\underline{d}_i^N - 1/\mathscr{X}}\right),
$$

where we denote $-\mathscr{A}^\gamma/(2\pi_1')$ by $\mathscr{X}$. If $\underline{d}_i^N = \omega(1)$, $((1 + 1/\mathscr{X} + 1/(2\mathscr{X}^2))^{\mathscr{X}})^{-\underline{d}_i^N - 1/\mathscr{X}} \to 0$, as $N \to \infty$. Hence, the event "$\mathcal{N}_i \geq 2, \mathcal{N}_i^{\mathrm{SD}} = 0|\mathscr{C}_{\mathbb{D}} = i$" holds *w.h.p.* when $\gamma \leq 2$, and at least with a constant probability $1 - \pi_1'$ when $\gamma > 2$, according to Lemma 3.1. If $\underline{d}_i^N = \Theta(1)$, $\mathbf{Pr}(\mathcal{N}_i \geq 2, \mathcal{N}_i^{\mathrm{SD}} = 0|\mathscr{C}_{\mathbb{D}} = i)$ is lower bounded by $(1 - \pi_1')(1 - e^{-\underline{d}_i^N})$. If $\underline{d}_i^N = o(1)$, $\mathbf{Pr}(\mathcal{N}_i \geq 2, \mathcal{N}_i^{\mathrm{SD}} = 0|\mathscr{C}_{\mathbb{D}} = i) \to 0$, as $N \to \infty$.

According to the results of partial sum of p-series, the probability of finding $\mathbb{S}$ and $\mathbb{D}$ on the same road segment during a slot is asymptotically given by

$$\sum_{i=1}^{G} \mathbf{Pr}(\mathscr{C}_{\mathrm{S}} = i | \mathscr{C}_{\mathrm{D}} = i)\pi_{\mathrm{D}}(i) = \sum_{i=1}^{G} \pi_{\mathrm{S}}(i)\pi_{\mathrm{D}}(i)$$

$$= \sum_{\alpha=1}^{\mathscr{A}} (16\alpha - 12)\alpha^{-2\gamma}\pi_1'^2 = \pi_1'^2 \sum_{\alpha=1}^{\mathscr{A}} \left(\frac{16}{\alpha^{2\gamma-1}} - \frac{12}{\alpha^{2\gamma}}\right) \qquad (3.12)$$

$$= \begin{cases} \Theta(\pi_1'^2 (N/\psi)^{\kappa(1-\gamma)}) & 0 < \gamma < 1 \\ \Theta(\pi_1'^2 \log(N/\psi)) & \gamma = 1 \\ \Theta(\pi_1'^2) & \gamma > 1. \end{cases}$$

Base on the analysis above, when $\underline{d}_i^N = \Omega(1)$, $\forall k$, w.h.p. we have,

$$\lambda_k(\Phi) \geq \frac{c' p_{ac}}{2\overline{d}_i^N} \sum_{i=1}^{G} \pi_{\mathrm{D}}(i) + \frac{p_{ac}}{\overline{d}_i^N} \sum_{i=1}^{G} \pi_{\mathrm{S}}(i)\pi_{\mathrm{D}}(i)$$

$$= c' p_{ac}/\overline{d}_i^N + O(1) p_{ac}/\overline{d}_i^N,$$

where c' is constant. Therefore, $\lambda(\Phi) = \Omega(p_{ac}/\overline{d}_i^N)$, w.h.p.. When $\underline{d}_i^N = o(1)$, $\forall k$,

$$\lambda_k(\Phi) \geq p_{ac}(\sum_{i=1}^{G} \pi_{\mathrm{S}}(i)\pi_{\mathrm{D}}(i))/\overline{d}_i^N.$$

Thus, w.h.p., we have

$$\lambda(\Phi) = \Omega(p_{ac}(\sum_{i=1}^{G} \pi_{\mathrm{S}}(i)\pi_{\mathrm{D}}(i))/\overline{d}_i^N).$$

From Lemmas 3.1 and 3.3, the assert follows.

We then derive an upper bound of per-vehicle throughput considering any possible stabilizing scheduling policies under $\mathscr{X}$-I. Let $\mathscr{X}_d(T)$ denote the number of packets delivered in the network through direct transmissions from the source to destination, and $\mathscr{X}_r(T)$ denote the number of packets delivered to the destination via relaying, during the interval $[0, T]$. Thus, provided the arbitrary and fixed $\epsilon > 0$, there must exist arbitrarily large values of T such that the per-vehicle throughput $\lambda(\Phi)$ satisfies

$$\frac{\mathscr{X}_d(T) + \mathscr{X}_r(T)}{T} \geq N\lambda(\Phi) - \epsilon. \qquad (3.13)$$

Let $\mathscr{Y}(T)$ denote the total number of transmission opportunities during the interval $[0, T]$. From (3.13), we have

$$\frac{1}{T}\mathscr{Y}(T) \geq \frac{1}{T}\mathscr{X}_d(T) + \frac{2}{T}\mathscr{X}_r(T)$$

$$\geq \frac{1}{T}\mathscr{X}_d(T) + 2\left((N\lambda(\Phi) - \epsilon) - \frac{1}{T}\mathscr{X}_d(T)\right).$$

The first inequality holds since the relayed packet reaches to the destination through at least two hops. Hence,

$$\lambda(\Phi) \leq \frac{\frac{1}{T}\mathscr{Y}(T) + \frac{1}{T}\mathscr{X}_d(T) + 2\epsilon}{2N},$$

i.e.,

$$\lambda(\Phi) \leq \lim_{T \to \infty} \frac{\frac{1}{T}\mathscr{Y}(T) + \frac{1}{T}\mathscr{X}_d(T)}{2N}. \tag{3.14}$$

Due to the transmission interference, the total number of transmission opportunities is no larger than the maximum number of concurrent transmissions during $[0, T]$. We have $\lim_{T \to \infty} \frac{1}{T}\mathscr{Y}(T) \leq Gp_{ac}$. Similarly, we have $\lim_{T \to \infty} \frac{1}{T}\mathscr{X}_d(T) \leq Gp_{ac}$, where the equality holds when there is always an $\mathbb{S}$-$\mathbb{D}$ transmission on each road segment of a non-interference group during each time slot. By plugging the inequalities into (3.14), we have

$$\lambda(\Phi) \leq \frac{Gp_{ac} + Gp_{ac}}{2N} = \frac{p_{ac}}{\psi} = \Theta\left(\frac{1}{\psi}\right). \tag{3.15}$$

From (3.15), the per-vehicle throughput capacity of Φ cannot be better than $\Theta\left(\frac{1}{\psi}\right)$.

3.3.3 Average Per-Vehicle Throughput

We next derive a lower bound of the average per-vehicle throughput $\tilde{\lambda}(\Phi)$, stated in Theorem 3.2, based on the proposed two-hop relay scheme for $\gamma \geq 2$, where the network shows dramatic social features. To simplify the analysis, we let $\psi = \Theta(1)$ in this subsection, i.e., the network density keeps constant and does not scale up with the population of vehicles. Considering all possible functions of ψ with the order of $o(N)$ makes the derivation very complex. The following lemmas (Lemmas 3.4–3.8) are first presented to prove Theorem 3.2.

Lemma 3.4. *Let $\mathcal{T}$ be a regular tessellation of the network, whose elements $\mathcal{T}_i$ contains $\lceil 100N^{1-\nu}\log(N)\rceil$ street blocks. w.h.p., every element of $\mathcal{T}$ contains at least one social spot.*

Proof. Recall that each element $\mathcal{T}_i$ of $\mathcal{T}$ is a regular area containing $\lceil 100N^{1-\nu}\log(N)\rceil$ street blocks. Let $\mathcal{T}_i^s$ denote the number of social spots contained in $\mathcal{T}_i$. Applying the Vapnik-Chervonenkis Theorem, $\forall \mathcal{T}_i$,

$$\mathbf{Pr}\left\{ \sup_{\mathcal{T}_i \in \mathcal{T}} \left| \frac{\mathcal{T}_i^s}{V} - \frac{100N^{1-\nu}\log(N)}{C} \right| \le \epsilon \right\} > 1 - \delta.$$

Note that VC-d($\mathcal{T}$) is at most 4. The condition (3.9) is satisfied when $\epsilon = \delta = \frac{50\log(N)}{N^\nu}$. Thus, $\forall \mathcal{T}_i$,

$$\mathbf{Pr}\{\mathcal{T}_i^s \ge 50\log(N)\} > 1 - \frac{50\log(N)}{N^\nu}.$$

The lemma follows as $N \to \infty$.

We denote by $P(\Phi) = E[\sum_{i=1}^{G} \mathbb{I}_{\mathcal{N}_i \ge 2}]$ the average number of road segments where there are at least two vehicles during a time slot. Similarly, let $Q(\Phi) = E[\sum_{i=1}^{G} \mathbb{I}_{\mathcal{N}_i^{\mathrm{SD}} \ge 1}]$ denote the average number of road segments where there is at least one S-D pair during a time slot. Lemmas 3.5 and 3.8 present a lower bound of $P(\Phi)$ respectively when $\nu \ne 1$ and $\nu = 1$.

Lemma 3.5. *When $\nu \ne 1$, w.h.p., we have,*

$$P(\Phi) = \begin{cases} \Omega(N^{\frac{2}{\gamma}+\nu(1-\frac{2}{\gamma})}/\log^3(N)) & \kappa + \nu > 1 \\ \Omega(N^{\nu+\frac{2\kappa}{\gamma}}/\log(N)) & \kappa + \nu < 1 \\ \Omega(N^{\nu+\frac{2\kappa}{\gamma+\vartheta}}/\log(N)) & \kappa + \nu = 1 \end{cases}$$

where ϑ is a positive and arbitrarily small value.

Proof. We consider a single social spot S_j in an area. Recall that S_j^ν is the number of vehicles associated with S_j. For road segment i in the mobility region of the vehicles, from (3.7), we have,

$$\mathbf{Pr}(\mathcal{N}_i \ge 2) = 1 - \mathbf{Pr}(\mathcal{N}_i \le 1)$$

$$\ge 1 - (1 - \pi'_{\mathcal{B}})^{\frac{(1-\varepsilon)N}{V}} - \frac{(1-\varepsilon)N}{V}\pi'_{\mathcal{B}}(1 - \pi'_{\mathcal{B}})^{\frac{(1-\varepsilon)N}{V}-1}$$

$$\ge 1 - e^{-\pi'_{\mathcal{B}}\frac{(1-\varepsilon)N}{V}} - \pi'_{\mathcal{B}}\frac{(1-\varepsilon)N}{V}e^{-\pi'_{\mathcal{B}}\frac{(1-\varepsilon)N}{V}+\pi'_{\mathcal{B}}}$$

$$= 1 - e^{-\pi'_{\mathcal{B}}\frac{(1-\varepsilon)N}{V}}(1 + \pi'_{\mathcal{B}}\frac{(1-\varepsilon)N}{V}e^{\pi'_{\mathcal{B}}})$$

where $\pi'_{\mathscr{B}} = \mathscr{B}^{-\gamma}\pi'_1$ and $\mathscr{B} \le \min\{\mathscr{A}, \lceil 10\sqrt{N^{1-\nu}\log(N)}\rceil\}$ from Lemma 3.4. Further, it is satisfied that $\pi'_{\mathscr{B}}\frac{N}{V} = \omega(1)$. Thus, letting $\varepsilon \to 0$, as $\to \infty$, the event "$\mathscr{N}_i \ge 2$" holds w.h.p.. Considering the regular tessellation $\mathscr{T}$ of the network, according to Lemma 3.4, w.h.p., we have,

$$P(\Phi) = \sum_{i=1}^{G} E[\mathbb{1}_{\mathscr{N}_i \ge 2}] = \sum_{i=1}^{G} \mathbf{Pr}(\mathscr{N}_i \ge 2)$$

$$\ge \frac{C}{100N^{1-\nu}\log(N)} \cdot \frac{1}{4} \cdot 4\mathscr{B}(2\mathscr{B} - 1) = \Theta(\frac{\mathscr{B}^2 N^\nu}{\log(N)}). \tag{3.16}$$

For $\kappa + \nu > 1$, we choose $\mathscr{B} = \Theta(N^{\frac{1-\nu}{\gamma}}/\log(N))$. $\mathscr{B}$ can scale as $\Theta(N^{\frac{\kappa}{\gamma+\vartheta}})$ for $\kappa + \nu = 1$. When $\kappa + \nu < 1$, $\mathscr{B}$ can be $\Theta(N^{\frac{\kappa}{\gamma}})$. The lemma follows.

Lemma 3.6. *(Chebyshev's Inequality) If X is a random variable with mean $E[X]$ and variance $Var(X)$, then, for any value $k > 0$,*

$$\mathbf{Pr}(|X - E[X]| \ge k) \le \frac{Var(X)}{k^2}.$$

Lemma 3.6 is well known and we use it to prove Lemma 3.7.

Lemma 3.7. *When $\nu = 1$, at least $(1 - e^{-\psi})C$ social spots associate with at least one S-D pair w.h.p..*

Proof. We denote by $\mathbb{I}_C = \sum_{i=1}^{C} \mathbb{1}_{S_i^\nu=0}$ the number of social spots that are not selected by any S-D pair in the network. $\mathbf{Pr}(\mathbb{1}_{S_i^\nu=0} = 1) = (1 - \frac{1}{C})^{\frac{N}{2}}$. Thus, the expectation and variance of $\mathbb{1}_{S_i^\nu=0}$ are $E[\mathbb{1}_{S_i^\nu=0}] = (1 - \frac{1}{C})^{\frac{N}{2}}$ and $Var(\mathbb{1}_{S_i^\nu=0}) = (1 - \frac{1}{C})^{\frac{N}{2}} - (1 - \frac{1}{C})^N$, respectively. Next we need to determine the variance of $\mathbb{I}_C$. For any $i \ne j$, $j \in \{1, 2, \ldots, C\}$, $Cov(\mathbb{1}_{S_i^\nu=0}, \mathbb{1}_{S_j^\nu=0}) = E[\mathbb{1}_{S_i^\nu=0}\mathbb{1}_{S_j^\nu=0}] - E[\mathbb{1}_{S_i^\nu=0}]E[\mathbb{1}_{S_j^\nu=0}]$, where $Cov(\mathbb{1}_{S_i^\nu=0}, \mathbb{1}_{S_j^\nu=0})$ is the covariance of variable $\mathbb{1}_{S_i^\nu=0}$ and $\mathbb{1}_{S_j^\nu=0}$. It is easy to get that $E[\mathbb{1}_{S_i^\nu=0}\mathbb{1}_{S_j^\nu=0}] = (1 - \frac{2}{C})^{\frac{N}{2}}$. Since $Cov(\mathbb{1}_{S_i^\nu=0}, \mathbb{1}_{S_i^\nu=0}) = Var(\mathbb{1}_{S_i^\nu=0})$, we have

$$Var(\mathbb{I}_C) = Var\left(\sum_{i=1}^{C} \mathbb{1}_{S_i^\nu=0}\right) = \sum_{i=1}^{C}\sum_{j=1}^{C} Cov(\mathbb{1}_{S_i^\nu=0}, \mathbb{1}_{S_j^\nu=0})$$

$$= \sum_{i=1}^{C} Cov(\mathbb{1}_{S_i^\nu=0}, \mathbb{1}_{S_i^\nu=0}) + 2\sum_{i=1}^{C}\sum_{j<i} Cov(\mathbb{1}_{S_i^\nu=0}, \mathbb{1}_{S_j^\nu=0})$$

$$= C\left((1 - \frac{1}{C})^{\frac{N}{2}} - (1 - \frac{1}{C})^N\right)$$

$$+ C(C-1)\left((1-\frac{2}{C})^{\frac{N}{2}} - (1-\frac{1}{C})^N\right)$$

$$\leq C\left((1-\frac{1}{C})^{\frac{N}{2}} - (1-\frac{1}{C})^N\right).$$

The inequality holds because $(1-\frac{2}{C})^{\frac{N}{2}} - (1-\frac{1}{C})^N = (1-\frac{2}{C})^{\frac{N}{2}} - (1-\frac{2}{C}+\frac{1}{C^2})^{\frac{N}{2}} \leq 0$.
From Lemma 3.6, choosing $k = \epsilon C$, we have

$$\mathbf{Pr}(\mathbb{I}_C - E[\mathbb{I}_C] \geq \epsilon C) \leq \frac{C[(1-\frac{1}{C})^{\frac{N}{2}} - (1-\frac{1}{C})^N]}{\epsilon^2 C^2}.$$

Note that $E[\mathbb{I}_C] = C(1-\frac{1}{C})^{\frac{N}{2}}$. Thus,

$$\mathbf{Pr}\left(\frac{\mathbb{I}_C}{C} \geq (\rho+\epsilon)\right) \leq \frac{\rho-\rho^2}{\epsilon^2} \cdot \frac{1}{C},$$

where $\rho = (1-\frac{1}{C})^{\frac{N}{2}}$. Since $N = 2\psi C$, as $N \to \infty$, $\rho \to e^{-\psi}$. Therefore, we have

$$\lim_{N\to\infty} \mathbf{Pr}(\mathbb{I}_C/C \geq e^{-\psi}) = 0,$$

i.e., the probability of $\mathbb{I}_C$ being over a constant proportion of C tends to zero as $N \to \infty$. The lemma follows.

Lemma 3.8. *When $v = 1$, w.h.p., $P(\Phi) = \Omega(N/\log^2(N))$ for $\gamma = 2$ and $\kappa \neq 0$; $P(\Phi) = \Theta(N)$ for $\gamma = 2$ and $\kappa = 0$; $P(\Phi) = \Theta(N)$ for $\gamma > 2$.*

Proof. According to Lemma 3.7, we can obtain that *w.h.p.*, there are at least $2(1 - e^{-\psi})C$ road segments, each of which belongs to a *Tier*(1) of vehicles' mobility region. Let $\mathscr{S}$ denote the set of road segments that are not contained in the mobility region of any vehicle. $\bar{\mathscr{S}}$ is the complementary set of $\mathscr{S}$ in $\{1, 2, \ldots, G\}$. Note that $\mathbf{Pr}(\mathscr{N}_i \geq 2) = 0, \forall i \in \mathscr{S}$. Thus,

$$P(\Phi) = \sum_{i=1}^{G} E[\mathbb{I}_{\mathscr{N}_i \geq 2}] = \sum_{i=1}^{G} \mathbf{Pr}(\mathscr{N}_i \geq 2)$$

$$= \sum_{i \in \mathscr{S}} \mathbf{Pr}(\mathscr{N}_i \geq 2) + \sum_{j \in \bar{\mathscr{S}}} \mathbf{Pr}(\mathscr{N}_j \geq 2)$$

$$= |\mathscr{S}| \cdot 0 + \sum_{j \in \bar{\mathscr{S}}} \mathbf{Pr}(\mathscr{N}_j \geq 2) \geq \sum_{j \in \bar{\mathscr{S}}} \pi_1'^2,$$

since for any $j \in \bar{\mathscr{S}}$, $\mathbf{Pr}(\mathscr{N}_j \geq 2) \geq \pi_1'^2$. The Lemma follows from Lemma 3.7.

Fig. 3.6 A decoupling queue structure

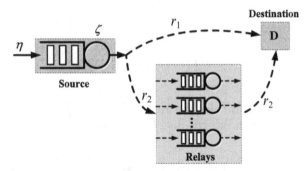

Theorem 3.2. *For the social-proximity grid-like VANETs, with the two-hop relay scheme $\mathscr{X}$, a bound of average per-vehicle throughput capacity $\tilde{\lambda}(\Phi)$ is given by w.h.p., (i) when $\nu \neq 1$ and $\gamma \geq 2$,*

$$
\tilde{\lambda}(\Phi) = \begin{cases}
\Omega(N^{\frac{2}{\gamma}+\nu(1-\frac{2}{\gamma})-1}/\log^3(N)) & \kappa+\nu>1 \\
\Omega(N^{\nu+\frac{2\kappa}{\gamma+\vartheta}-1}/\log(N)) & \kappa+\nu=1 \\
\Omega(N^{\nu+\frac{2\kappa}{\gamma}-1}/\log(N)) & \kappa+\nu<1
\end{cases}
$$

and (ii) when $\nu = 1$, $\tilde{\lambda}(\Phi) = \Omega(\frac{1}{\log^2(N)})$, for $\gamma = 2$ and $\kappa \neq 0$; $\tilde{\lambda}(\Phi) = \Theta(1)$ for $\gamma = 2$ and $\kappa = 0$; $\tilde{\lambda}(\Phi) = \Theta(1)$ for $\gamma > 2$.

Proof. Under the two-hop relay scheme $\mathscr{X}$, we can use a decoupling queue structure, similar to that in [84], to model each unicast flow, as shown in Fig. 3.6. Without loss of generality, we consider that the packet arrival rate η follows the Bernoulli process. In other words, in each unicast flow, one packet arrives with the probability η at the current slot, and with the rest probability if there is no packet arrival. Hence, the source vehicle, e.g., v_k, can be represented as a Bernoulli/Bernoulli queue with packet arrival rate η_k and service rate ζ_k. The buffering packet in the source is transmitted (served) to either its destination directly or one of the relays within the mobility region of the source. The transmission opportunity arises with probability ζ_k. Let $r_1^k(N)$ denote the long term average rate at which a direct transmission to the destination is scheduled to source v_k, and $r_2^k(N)$ denote the long term average rate at which a source-to-relay transmission is scheduled to source v_k. The transmission opportunity arises at the rate $\zeta_k(N) = r_1^k(N) + r_2^k(N)$. As per the definition, $\tilde{\lambda}(\Phi) = \frac{\sum_{k=1}^{N}\zeta_k(N)}{N}$. Since the two-hop relay scheme $\mathscr{X}$ schedules a source-to-relay transmission and a relay-to-destination transmission equally likely, the rate into the relays is equal to the rate out of the relays. During each time slot, the total number of transmission opportunities over the network is $\sum_{k=1}^{N}(r_1^k(N) + 2r_2^k(N))$. Given that the transmission opportunity arises on a road segment when it is active and at least two vehicles are on it, we have,

$$p_{ac} P(\Phi) = \sum_{k=1}^{N} (r_1^k(N) + 2r_2^k(N)). \tag{3.17}$$

Since the two-hop relay scheme $\mathscr{X}$ schedules the source-to-destination transmission whenever possible, we have,

$$p_{ac} Q(\Phi) = \sum_{k=1}^{N} r_1^k(N). \tag{3.18}$$

From (3.17) and (3.18), we obtain $\sum_{k=1}^{N} r_2^k(N) = \frac{p_{ac}(P(\Phi)-Q(\Phi))}{2}$ and therefore

$$\tilde{\lambda}(\Phi) = \frac{\sum_{k=1}^{N} (r_1^k(N) + r_2^k(N))}{N} = \frac{p_{ac}(P(\Phi) + Q(\Phi))}{2N}.$$

Since $P(\Phi) \geq Q(\Phi)$ and from Lemmas 3.5 and 3.8, the theorem follows.

Remark. The average per-vehicle throughput is analyzed as a global metric to evaluate the network performance with inhomogeneous vehicle density. For example, from Theorem 3.2, we can attain that the constant per-vehicle throughput is feasible w.h.p. for N_f S-D pairs, where $N_f = \Theta(N) \leq \frac{N}{2}$, when $\nu = 1$ and $\gamma > 2$. Due to the socialized mobility of vehicles and the randomness of the locations of vehicle's social spots, the network shows spatial variations of vehicle density. Hence, the throughput performance of vehicles in different areas of the city grid may be different. For example, in a hot area where is covered by a large number of different overlapped mobility regions, the throughput of an S-D pair in that area may drop significantly.

3.4 Case Study

We present a case study for some specific values of γ and ψ. A graphical representation of our results is reported in Figs. 3.8 and 3.9. The results are shown in log-scale in terms of κ and ν, with $\psi = \Theta(1)$. For example, "-0.5" represents a throughput of $\Theta(\frac{1}{\sqrt{N}})$. Our analytical results demonstrate three possible regimes depending on different values of κ and ν, as shown in Fig. 3.7. Recall that the size of mobility region and the number of social spots scale as $\Theta((N/\psi)^{\kappa-1})$ and $\Theta((N/\psi)^{\nu})$, respectively.

(i) **Dense regime**: when $\kappa + \nu > 1$, the sum of all mobility regions associated with different social spots is $\Theta((N/\psi)^{\kappa+\nu-1}) = \omega(1)$, which indicates that V different mobility regions are overlapped and fully cover the city grid;

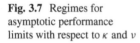

Fig. 3.7 Regimes for
asymptotic performance
limits with respect to κ and v

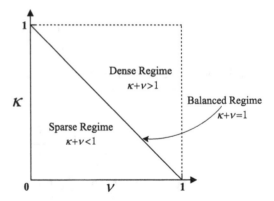

(ii) **Sparse regime**: when $\kappa + v < 1$, the sum of all different mobility regions is
$o(1)$, which results in V typically isolated mobility regions sparsely distributed
in the city grid; and

(iii) **Balanced regime**: when $\kappa + v = 1$, the sum of all different mobility regions
has the same scale with the grid area, making the mobility area of vehicles
perfectly fit the grid area in the order sense.

The lower bound of the per-vehicle throughput capacity for $\gamma = 2$ is shown
in Fig. 3.8. In the dense regime, bounds of per-vehicle throughput capacity are
dominated by κ given γ and ψ. Notably, a large κ indicates a large size of mobility
region, which results in decrease in per-vehicle throughput due to the following
reasons. (1) the contact probability of a particular pair of vehicles is reduced; and
(2) different mobility regions are largely overlapped so that the vehicle density is
potentially increased. In the sparse regime, the performance is mostly dominated by
v. When v tends to 1, the number of vehicles associated with each social spot is
dramatically reduced, avoiding a high vehicle density in the proximity of social
spots. Hence, the throughput performance is enhanced with a large value of v.
The performance decreases in the sparse regime when $\kappa + v$ tends to zero, due to
increasing empty area in the city grid where there is no any packet transmission
occurs. When $\kappa + v = 1$, the network achieves optimal bounds of per-vehicle
throughput capacity, since the geographic area of the city grid, i.e., the spatial
resource of the network, is just fully utilized for packet transmissions. Thus, it
is possible to achieve almost constant (except for the polylogarithmic factor) per-
vehicle throughput, i.e., in the case of $\kappa = 0$ and $v = 1$.

The average per-vehicle throughput is a global performance metric for the
network with inhomogeneous vehicle densities. Figure 3.9a, b demonstrate the
average per-vehicle throughput for $\gamma = 2$ and $\gamma = 4$, respectively. When $\gamma = 2$,
almost constant average per-vehicle throughput is achievable with high probability
in the dense regime. However, in this case, the per-vehicle throughput may degrade
significantly, as shown in Fig. 3.8, in some hot area where is covered by a large
number of different overlapped mobility regions. With a larger value of γ, e.g.,
$\gamma = 4$, vehicles usually move in a very limited area centered at the social spot.

Fig. 3.8 Lower bound of
per-vehicle throughput

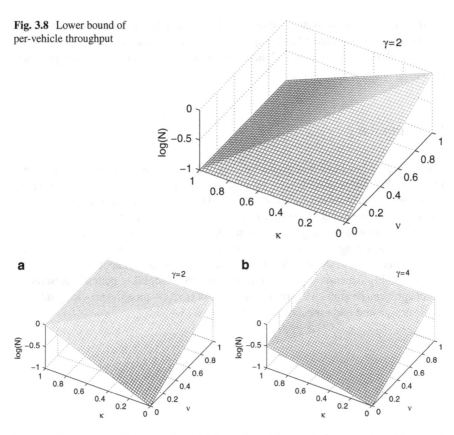

Fig. 3.9 Average per-vehicle throughput. (**a**) Lower bound for $\gamma = 2$. (**b**) Lower bound for $\gamma = 4$

Due to the limited spatial resource, the total number of concurrent transmissions
is reduced. As a result, the average per-vehicle throughput decreases since the
geographic area of the city grid is not fully used for packet transmissions.

The analysis in this chapter shows that the throughput capacity depends on
inherent mobility patterns of the network. Notably, the parameters of the socialized mobility are possible to obtain, e.g., they can be extracted from real-world
mobility traces of vehicles. Once the mobility pattern of the real-world scenario
is determined, our results can be applied to predict network performance, at least
in the order sense. We provide an example in the following. Consider a network
of 10^4 vehicles with parameters of mobility model $\kappa = 0.5$, $\nu = 0.7$, and
$\gamma = 2$. The bandwidth of point-to-point link is 1 Mbps. From the results in this
chapter, neglecting polylogarithmic factors and constant factor ψ, a per-vehicle
throughput of around 10 kbps is achievable. Considering another mobility pattern
where $\kappa = 0.2$ and $\nu = 0.8$, the per-vehicle throughput is around 150 kbps. From
reference [29], we know that in fact there exists a throughput-delay tradeoff for
a given mobility pattern. For example, for $\kappa = 0.5$, $\nu = 0.7$, the network delay

could be from seconds to days under two-hop relay scheme; whereas better delay performance can be obtained by using other forwarding schemes, such as multi-hop scheme with or without packet redundancies, however, with a lower throughput than 10 kbps.

3.5 Summary

In this chapter, we have investigated the asymptotic capacity for social-proximity VANETs. We adopt a scalable city grid to deploy the VANET and consider a socialized mobility model for each vehicle. The user applications have proximity nature, i.e., the source and the destination of each flow have the same social spot. Under the proposed two-hop relay scheme, the bounds of the per-vehicle throughput capacity and average per-vehicle throughput have been derived with respect to different network parameters. We have shown that the throughput capacity of the network highly depend on the inherent parameters of mobility patterns. Results in this chapter can be applied to predict the network performance and provide guidance on the design and implementations for large-scale VANETs.

Chapter 4
Downlink Capacity of Vehicular Networks with Access Infrastructure

Wireless access infrastructure, such as Wi-Fi access points and cellular base stations, plays a vital role in offering pervasive Internet services to vehicles. However, the deployment costs of different access infrastructure are highly variable. In this chapter, we analyze the downlink capacity of vehicles and investigate the capacity-cost tradeoffs for the network in which access infrastructure is deployed to provide a downlink data pipe to all vehicles. Three alternatives of wireless access infrastructure are considered, i.e., cellular base stations (BSs), wireless mesh backbones (WMBs), and roadside access points (RAPs). We first derive a lower bound of downlink capacity for each type of access infrastructure. We then present a case study based on a ideal city grid of $400\,km^2$ with 0.4 million vehicles, in which we examine the capacity-cost tradeoffs for different deployment solutions in terms of both capital expenditures (CAPEX) and operational expenditures (OPEX). Rich implications from the results provide fundamental guidance on the choice of cost-effective wireless access infrastructure for the emerging vehicular networking.

4.1 Introduction

Three candidate solutions are considered to provide Internet connectivity to vehicles, i.e., off-the-shelf 3 G or 4 G cellular networks, drive-thru or roadside Wi-Fi access points, and the wireless mesh backbone, which consists of wirelessly connected mesh nodes (MNs) including one gateway to the Internet. Since VANETs have yet to become reality, there remains great uncertainty as to the feasibility of each type of wireless access infrastructure in terms of both network performance and deployment cost.

In this chapter, to better understand the capacity-cost issue in vehicular access networks, we consider a scalable urban area where vehicles access Internet through deployed infrastructure nodes. We first analyze the downlink capacity of vehicles to show how it scales with the number of deployed infrastructure nodes.

N. Lu and X. Shen, *Capacity Analysis of Vehicular Communication Networks*, SpringerBriefs in Electrical and Computer Engineering, DOI 10.1007/978-1-4614-8397-7__4, © The Author(s) 2014

The downlink capacity is defined as the maximum average downlink throughput achieved *uniformly* by all the vehicles from the access infrastructure. Two operation modes of the network are considered to provide pervasive Internet access: *infrastructure mode*, in which the network is fully covered by infrastructure nodes, i.e., all the vehicles are within the coverage of the infrastructure, and thereby only the infrastructure-to-vehicle (I2V) communication is utilized to deliver the downlink traffic; and *hybrid mode*, in which the network is not fully covered and the downlink flow is relayed to the vehicles in the area without coverage by means of *multi-hop* vehicle-to-vehicle (V2V) communications. A lower bound of the downlink capacity is derived for the network with deployment of cellular base stations (BSs), wireless mesh backbones (WMBs), and roadside access points (RAPs), respectively. To understand the effect of key factors, such as the deployment scale and the coverage of infrastructure nodes, we present a case study based on a ideal city grid of $400 \, \text{km}^2$ with 0.4 million vehicles. Furthermore, we examine the capacity-cost tradeoffs for different deployments. We show that in the hybrid mode, to achieve the same downlink throughput, the network roughly needs X BSs, or $6X$ MNs, or $25X$ RAPs[1]; whereas in the infrastructure mode, if it is desired to improve the downlink throughput by the same amount for each deployment, we roughly need to additionally deploy X BSs, or $5X$ MNs, or $1.5X$ RAPs. By explicitly taking capital expenditures (CAPEX) and operational expenditures (OPEX) of access infrastructures into consideration, the deployment of BSs or WMBs is cost-effective to offer a low-speed downlink rate to vehicles; nonetheless, when providing a high-speed Internet access, the deployment of RAPs outperforms the other two alternatives in terms of deployment costs. This implication can provide valuable guidance on the choice of access infrastructures for the automobile and telecommunication industry. In particular, as automotive industry gears for supporting high-bandwidth applications, non-cellular access infrastructure will play an increasingly important role in offering a cost-effective data pipe for vehicles.

To the best of our knowledge, this research represents the first theoretical study on the capacity-cost tradeoffs when providing pervasive Internet access to vehicles. [128] is the most relevant literature, in which Banerjee *et al.* first examined the performance-cost tradeoffs for VANETs by considering three infrastructure enhancement alternatives: BSs, meshes, and relays. They showed that if the average packet delay can be reduced by a factor of two by adding X BSs, the same reduction needs $2X$ MNs or $5X$ relays. They argued that relays or meshes can be a more cost-effective enhancement due to the high cost of deploying BSs. The objective of their work is to improve network delay by augmenting mobile ad hoc networks with infrastructure, which is different from ours. In addition, our methodology is also different from that adopted in [128]. Notably, quite a few research works [129–131] focus on content downloading in VANETs. Although we consider a downlink scenario as well, our focus is to unveil capacity-cost tradeoffs for deployment of vehicular access networks.

---

[1] X is used to represent a ratio relationship rather that a specific value.

The remainder of this chapter is organized as follows: Sect. 4.2 introduces the system model. In Sect. 4.3, we analyze the downlink capacity for each type of infrastructure deployment. We present the case study and examine the capacity-cost tradeoffs in Sect. 4.4. Section 4.5 summarizes this chapter.

4.2 System Model

4.2.1 Urban Street Pattern

Similar to Chap. 3, the street layout of the urban area is modeled by a perfect grid $\mathbb{G}(M, L)$, which consists of a set of M vertical roads intersected with a set of M horizontal roads. Each line segment of length L represents a road segment, as shown in Fig. 4.1. Let $\mathbb{G}$ be a torus to eliminate the border effects. We denote the total number of road segments in $\mathbb{G}$ by $\mathscr{G} = 2(M - 1)^2$. The scale of the urban grid is therefore determined by M and L. For example, M is roughly 100 and L is generally from 80 to 200 m for the downtown area of Toronto [121]. A summary of the mathematical notations used in the chapter is given in Table 4.1.

4.2.2 Spatial Distribution of Vehicles

Taking a snapshot of the city grid where vehicles are moving, it is considered that vehicles are distributed according to a Poisson Point Process (p.p.) $\varPhi$ with intensity measure $\varXi$ on $\mathbb{G}(M, L)$. Further, $\varXi(\mathrm{d}x) = \xi \mathrm{d}x$, where $\xi \in (0, +\infty)$, indicating that the average number of vehicles on the road of length $\mathrm{d}x$ is $\xi \mathrm{d}x$. We denote by N the average number of vehicles in the grid. Thus,

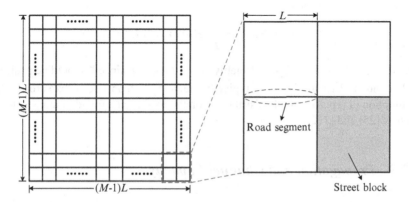

Fig. 4.1 A grid-like urban street pattern

Table 4.1 The notations for Chap. 4

Symbol	Description
N	The average number of vehicles in the grid
M	The number of parallel roads in the grid
L	The length of road segment
$\mathscr{G}$	The total number of road segments
$\mathbb{G}$	The urban grid
β	Path-loss exponent
ξ	Vehicle density
W	Communication bandwidth
θ	Ratio between the number of MGs and N_M
N_B	The number of deployed BSs
N_M	The number of deployed MNs
N_R	The number of deployed RAPs
R_V	Transmission radius of V2V communications
R_M	Transmission radius of M2M communications
$\tau_{\mathscr{B}}$	The number of tiers in BS service square
$\tau_{\mathscr{C}}$	The number of tiers in the coverage of BS
λ_B	Downlink capacity for deployment of BSs
λ_B^P	Downlink capacity of B2V transmissions
λ_B^A	Downlink capacity of V2V transmissions (BS)
τ_M	The number of tiers in WMB service square
τ_{MR}	The number of tiers in the coverage of MN
τ_W	The number of tiers in the coverage of WMB
λ_M	Downlink capacity for deployment of WMBs
λ_M^M	Downlink capacity of M2M transmissions
λ_M^P	Downlink capacity of M2V transmissions
λ_M^A	Downlink capacity of V2V transmissions (WMB)
L_R	Service region of an RAP
R_C	Transmission radius of RAP
λ_R	Downlink capacity for deployment of RAPs
λ_R^P	Downlink capacity of R2V transmissions
λ_R^A	Downlink capacity of V2V transmissions (RAP)

$$N = \Xi(\mathbb{G}) = \int_{\mathbb{G}} \Xi(\mathrm{d}x) = \mathscr{G}L\xi. \qquad (4.1)$$

And then $\xi = \frac{N}{\mathscr{G}L} = \frac{N}{2L(M-1)^2}$. Notably, $M = \Theta(\sqrt{N})$, since ξ should be positive and bounded. Particularly, ξL is typically much larger than 1 for urban areas. The assumption of p.p. for vehicle distribution on the road can be found in many studies such as [25] and [132].

4.2.3 Propagation and Channel Capacity

The received signal power P_{ij} at receiver j from transmitter i follows the propagation model described in the following: $P_{ij} = KP_i/l(d_{ij})$, where P_i is the transmission power of transmitter i, d_{ij} is the Euclidean distance between i and

j, and K is a parameter related to the hardware of communication systems. The path-loss function is given by $l(d_{ij}) = (d_{ij})^\beta$, where β is positive and called the path-loss exponent. Typically, we have $\beta = 4$ for urban environments [133]. Note that the phenomenon of channel fluctuations is not considered since a macroscopic description of power attenuation shown above is sufficient for throughput analysis of a long-term average.

The channel capacity of transmitter i and its receiver j is described by Shannon capacity:

$$\mathcal{T}_{ij} = W_{ij} \log_2(1 + SINR_{ij}), \tag{4.2}$$

where W_{ij} is the spectrum bandwidth for the transmission and $SINR_{ij}$ is the *signal-to-interference-plus-noise ratio* (SINR) at receiver j. The interference experienced by receiver j is the aggregation of the signal powers received from all simultaneous transmitters, except its own transmitter i. For ease of comparison, the same path-loss exponent and total bandwidth, which is denoted by W, are applied for each type of deployment of wireless access infrastructure.

4.3 Analysis of Downlink Capacity

We derive a lower bound of downlink capacity in this section for each type of infrastructure deployment, i.e., BSs, WMBs, and RAPs. Asymptotic results are also provided, describing how the downlink capacity scales with the number of deployed infrastructure nodes (directly related to the deployment cost). The derivations in this chapter are mainly based on geometric considerations about interference patterns under certain bandwidth planning. Note that the coverage of the infrastructure node is treated independently from the transmission power in the analysis. It is not necessary to explicitly show the relationship between these two parameters, since our results only depend on the coverage of infrastructure node. In addition, it is noteworthy that the difference between WMB and RAP is that WMBs use wireless mesh-to-mesh links as backhaul, whereas RAPs fully rely on external wired connectivity.

4.3.1 Network with Deployment of BSs

Let N_B denote the number of BSs deployed in the city grid $\mathbb{G}(M, L)$. The grid is thereby divided into N_B squares of equal area, which is denoted by $\mathcal{B}$ and hence $|\mathcal{B}| = (M - 1)^2 L^2 / N_B$. Each square is associated with one BS, which is deployed in the central street block of the square, as shown in Fig. 4.2. It is required that $N_B < (M - 1)^2$, i.e., the number of deployed BSs should be less than the total number of street blocks of $\mathbb{G}$. Furthermore, each square consists of multiple tiers

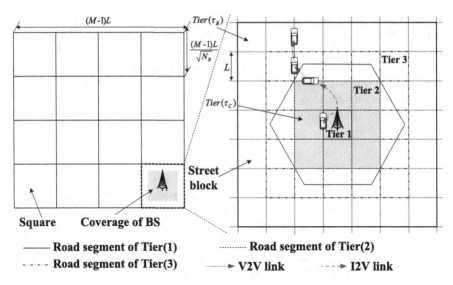

Fig. 4.2 Grid-like VANETs with deployment of cellular BSs

co-centered at the BS. $Tier(1)$ of the square is the street block where the BS is located and contains four road segments. The adjacent street blocks surrounding $Tier(1)$ form $Tier(2)$, and so on. It is easy to find that $Tier(\tau)$ contains $16\tau - 12$ road segments. We denote by $\tau_{\mathscr{B}}$ the number of tiers of each square. Hence,

$$\tau_{\mathscr{B}} \leq \lceil \frac{1}{2}\sqrt{\frac{|\mathscr{B}|}{L^2}} + 1 \rceil = \lceil \frac{M-1}{2\sqrt{N_B}} + 1 \rceil, \tag{4.3}$$

where $\lceil \cdot \rceil$ is the ceiling function.

The coverage of the BS is simply considered as a square area of $\tau_{\mathscr{C}}$ tiers, although it is often assumed that the cellular BS covers a hexagon region. A similar approximation can be found in [134]. When $\tau_{\mathscr{C}} \geq \tau_{\mathscr{B}}$, we let $\tau_{\mathscr{C}} = \tau_{\mathscr{B}}$. In this case, the network is fully covered by BSs and thereby operates in the infrastructure mode. When $\tau_{\mathscr{C}} < \tau_{\mathscr{B}}$, the network is partially covered by BSs and operates in the hybrid mode, i.e., BS-to-vehicle (B2V) transmissions and vehicle-to-vehicle (V2V) transmissions coexist. Let $\lambda_B(N, N_B)$ denote the downlink capacity for the deployment of BSs. Furthermore, we denote by λ_B^P and λ_B^A the downlink capacity of B2V and V2V transmissions, respectively. The downlink capacity in the hybrid mode is determined in the following.

$$\lambda_B(N, N_B) = \min \{\lambda_B^P, \lambda_B^A\}. \tag{4.4}$$

We first study the downlink throughput λ_B^P for B2V transmissions in the hybrid mode. The total bandwidth W is further divided into αW and $(1-\alpha)W$ respectively for B2V and V2V transmissions. A simple spectrum reuse scheme is adopted to mitigate the interference from neighboring squares in B2V transmissions: a square and its eight neighboring squares use different channels for B2V transmissions, each of which is of bandwidth $\alpha W/9$.

Let P_r^τ denote the received signal power of vehicle $\mathscr{V}_0$ on a road segment of $Tier(\tau)$ from its own BS in the square $\mathscr{S}_0$, where $\tau \leq \tau_{\mathscr{C}}$. From the propagation model, we have

$$P_r^\tau \geq \frac{KP_B}{\left[\sqrt{2}L(\tau - \frac{1}{2})\right]^\beta},\tag{4.5}$$

where P_B is the transmission power of BSs. The interference experienced by $\mathscr{V}_0$, denoted by I_B, is the aggregated signal power of all the other BSs transmitting on the same channel. Thus

$$I_B \leq \sum_{q=1}^{\infty} 8q \cdot \frac{KP_B}{\left[(3q - \frac{1}{2})\sqrt{|\mathscr{B}|}\right]^\beta} = \sum_{q=1}^{\infty} \frac{8qKP_B}{\left[(3q - \frac{1}{2})\frac{(M-1)L}{\sqrt{N_B}}\right]^\beta}$$

$$\leq \frac{8KP_B N_B^{\frac{\beta}{2}}}{L^\beta(M-1)^\beta}\left[\left(\frac{2}{5}\right)^\beta + \int_1^\infty \frac{1}{(3q-\frac{1}{2})^{\beta-1}}dq\right]$$

$$\leq \frac{2^{\beta+1}KP_B N_B^{\frac{\beta}{2}}}{5^\beta L^\beta(M-1)^\beta} \cdot \frac{12\beta+1}{3\beta-6}.$$

Given that $\mathscr{V}_0$ is on a road segment of $Tier(\tau)$, the SINR of the received signal from the BS at $\mathscr{V}_0$ is given by

$$SINR_\tau \geq \frac{5^\beta(3\beta-6)}{(12\beta+1)2^{\frac{3}{2}\beta+1}}\left[\frac{M-1}{(\tau-\frac{1}{2})\sqrt{N_B}}\right]^\beta.\tag{4.6}$$

Throughout this chapter, we neglect the noise as did in previous works like [111] and [112], since we focus on an interference-dominated vehicular environment.

For $\mathscr{V}_0$ on a road segment of $Tier(\tau)$, where $\tau \leq \tau_{\mathscr{C}} - 1$, from (4.2), we have

$$\lambda_B^P = W_\tau \log_2(1 + SINR_\tau),\tag{4.7}$$

where W_τ out of $\alpha W/9$ is the bandwidth allocated to a single vehicle on a road segment of $Tier(\tau)$. Since vehicles on road segments of $Tier(\tau_{\mathscr{C}})$ are required to relay the downlink traffic to vehicles in the area without the BS coverage (see Fig. 4.2), we have

$$\lambda_B^P = \frac{W_{\tau_{\mathscr{C}}} \log_2(1 + SINR_{\tau_{\mathscr{C}}})}{\left(\sum_{\tau=\tau_{\mathscr{C}}}^{\tau_{\mathscr{B}}} 16\tau - 12\right)/(16\tau_{\mathscr{C}} - 12)}.\tag{4.8}$$

From (4.7) and (4.8), we have

$$\sum_{\tau=1}^{\tau_{\mathscr{C}}-1} \frac{(16\tau - 12)\xi L\lambda_B^P}{\log_2(1 + SINR_\tau)} + \frac{(\sum_{\tau=\tau_{\mathscr{C}}}^{\tau_{\mathscr{B}}} 16\tau - 12)\xi L\lambda_B^P}{\log_2(1 + SINR_{\tau_{\mathscr{C}}})} = \frac{\alpha W}{9}.$$

Hence, $\lambda_B^P = \frac{\alpha W/9}{\xi L \mathscr{U}_1}$, where

$$\mathscr{U}_1 = \sum_{\tau=1}^{\tau_{\mathscr{C}}-1} \frac{16\tau - 12}{\log_2(1 + SINR_\tau)} + \frac{\sum_{\tau=\tau_{\mathscr{C}}}^{\tau_{\mathscr{B}}} 16\tau - 12}{\log_2(1 + SINR_{\tau_{\mathscr{C}}})}$$

$$\leq \frac{\sum_{\tau=1}^{\tau_{\mathscr{B}}} 16\tau - 12}{\log_2(1 + SINR_{\tau_{\mathscr{C}}})} = \frac{4\tau_{\mathscr{B}}(2\tau_{\mathscr{B}} - 1)}{\log_2\left(1 + \mathscr{U}_2\left[\frac{M-1}{(\tau_{\mathscr{C}}-\frac{1}{2})\sqrt{N_B}}\right]^\beta\right)}$$

$$\leq \frac{2(\frac{M-1}{\sqrt{N_B}} + 4)^2}{\log_2\left(1 + \mathscr{U}_2\left[\frac{M-1}{(\tau_{\mathscr{C}}-\frac{1}{2})\sqrt{N_B}}\right]^\beta\right)}.$$

The inequalities hold according to (4.3) and (4.6). We denote $\frac{5^\beta(3\beta-6)}{(12\beta+1)2^{\frac{3}{2}\beta+1}}$ by $\mathscr{U}_2$. A lower bound of λ_B^P is given by

$$\lambda_B^P \geq \frac{\alpha W/(9\xi L)}{2(\frac{M-1}{\sqrt{N_B}} + 4)^2} \log_2\left(1 + \mathscr{U}_2\left[\frac{M-1}{(\tau_{\mathscr{C}} - \frac{1}{2})\sqrt{N_B}}\right]^\beta\right). \tag{4.9}$$

We denote $\tau_{\mathscr{C}} = \tau_{\mathscr{B}}^\kappa$, $0 < \kappa < 1$ and $N_B = N^\nu$, $0 < \nu < 1$. Asymptotically, it is clear that $\lambda_B^P = \Omega(\frac{N_B}{N} \log_2(\frac{N}{N_B})) = \Omega(N^{\nu-1} \log_2 N)$. Notably, $\lambda_B^P = \Omega(\frac{N_B}{N}) = \Omega(N^{\nu-1})$ when $\kappa = 1$, i.e., the network operates in the infrastructure mode.

We then study the downlink capacity λ_B^A for V2V transmissions. Let P_V and $R_V(\geq L)$ denote the transmission power and the transmission radius of V2V communications, respectively. The Carrier Sensing Multiple Access (CSMA) with a carrier sensing radius of $2R_V$ is applied by vehicles to access the channel of bandwidth $(1 - \alpha)W$. Due to that simultaneous transmitters cannot be within a distance of $2R_V$ as per the stipulation of CSMA, the distribution of transmitting vehicles in the area without the BS coverage follows a Matérn-like hard core (MHC) p.p. [135]. The MHC p.p. is a dependent marked p.p. of original Poisson p.p. Φ of vehicles. Following [136], an average medium access probability over all the vehicles of Φ is given by

$$P_{ac} = (1 - e^{-\bar{\mathscr{N}}})/\bar{\mathscr{N}},$$

Fig. 4.3 A triangular lattice
of simultaneous transmitting
vehicles

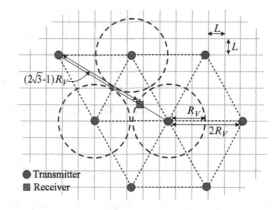

where $\bar{\mathcal{N}}$ is the average number of neighbors of a generic vehicle within the carrier
sensing range. We have

$$\bar{\mathcal{N}} \leq \xi L \cdot 2 \lceil \frac{4R_V}{L} \rceil (\lceil \frac{4R_V}{L} \rceil + 1)$$

$$\leq 8\xi L \left(\frac{2R_V}{L} + 1 \right)^2.$$

Therefore,

$$P_{ac} \geq \frac{1 - \exp\left(-8\xi L(2R_V/L + 1)^2\right)}{8\xi L(2R_V/L + 1)^2}. \tag{4.10}$$

Since $\exp\left(-8\xi L(2R_V/L + 1)^2\right)$ decays to 0 very fast, we can ignore this exponential term in (4.10).

In V2V transmissions, the received signal power at destination $\mathcal{V}_0$ from its transmitter is given by $P_r \geq KP_V/R_V^\beta$. Let $I_{\mathcal{V}_0}$ denote the aggregate interference power experienced by $\mathcal{V}_0$. A close-form expression of $I_{\mathcal{V}_0}$ is difficult to determine. Therefore, we derive an upper bound of $I_{\mathcal{V}_0}$ in the following. Since we consider a high density urban environment, simultaneous V2V transmitters under CSMA scheme with carrier sensing radius $2R_V$ cannot be denser than a triangular lattice [137]. As shown in Fig. 4.3, the six nearest interferers in the first layer are at distance $2R_V$. The next 12 form the second layer, and so on. The distance between the receiver marked and interferers in the first layer is at least R_V, and at least $(\sqrt{3}q - 1)R_V$ in the qth layer, $q > 1$. Thus,

$$I_{\mathcal{V}_0} \leq \frac{6KP_V}{R_V^\beta} + \sum_{q=2}^{\infty} 6q \cdot \frac{KP_V}{\left[(\sqrt{3}q - 1)R_V\right]^\beta}$$

$$\leq \frac{6KP_V}{R_V^\beta}\left[1 + \int_1^\infty \frac{1}{(\sqrt{3}q-1)^{\beta-1}}dq\right]$$

$$= \frac{6KP_V}{R_V^\beta}\left(1 + \frac{1}{\sqrt{3}(\beta-2)(\sqrt{3}-1)^{\beta-2}}\right).$$

We denote by $SINR_V$ the SINR of received signal at $\mathscr{V}_0$ from its V2V transmitter. It follows that

$$SINR_V \geq \frac{(\beta-2)(\sqrt{3}-1)^{\beta-2}}{2\sqrt{3}+(\beta-2)(\sqrt{3}-1)^{\beta-2}} = \mathscr{U}_3(\beta). \qquad (4.11)$$

Notably, $SINR_V$ is lower bounded by $\mathscr{U}_3(\beta)$, which merely depends on β.

Note that vehicles on road segments of $Tier(\tau_\mathscr{C})$ are required to relay the downlink traffic to vehicles from $Tier(\tau_\mathscr{C}+1)$ to $Tier(\tau_\mathscr{B})$. On the average, every vehicle on road segments of $Tier(\tau_\mathscr{C})$ needs to relay the traffic for $\bar{\eta}_1$ vehicles. We have,

$$
\begin{aligned}
\bar{\eta}_1 &= \frac{(\sum_{\tau=\tau_\mathscr{C}+1}^{\tau_\mathscr{B}} 16\tau - 12)\xi L}{(16\tau_\mathscr{C}-12)\xi L} \\
&= \frac{(2\tau_\mathscr{B}+2\tau_\mathscr{C}-1)(\tau_\mathscr{B}-\tau_\mathscr{C})}{4\tau_\mathscr{C}-3} \sim \frac{\tau_\mathscr{B}^{2-\kappa}-\tau_\mathscr{B}^\kappa}{2}.
\end{aligned}
\qquad (4.12)
$$

Recall that $\tau_\mathscr{C} = \tau_\mathscr{B}^\kappa$, $0 < \kappa < 1$. Hence, from (4.10) to (4.12), the downlink capacity λ_B^A can be lower bounded as follows.

$$
\begin{aligned}
\lambda_B^A &\geq \frac{(1-\alpha)W\log_2(1+SINR_V)P_{ac}}{\bar{\eta}_1} \\
&\geq \frac{(1-\alpha)W\log_2(1+\mathscr{U}_3(\beta))}{8\xi L(2R_V/L+1)^2\bar{\eta}_1} \\
&\sim \frac{(1-\alpha)W\log_2(1+\mathscr{U}_3(\beta))}{4\xi L(2R_V/L+1)^2 \cdot (\frac{M-1}{2\sqrt{N_B}}+2)^{2-\kappa}}.
\end{aligned}
\qquad (4.13)
$$

Let $(R_V/L) = \tau_\mathscr{B}^\mu$ establish a relationship between the transmission range of vehicles and the number of tiers of $\mathscr{B}$, where $0 < \mu < 1$. Further, it is required that $\mu < \kappa$, due to that the transmission range of vehicles should be smaller than that of BSs. Then, we obtain an asymptotic lower bound of λ_B^A from (4.13), i.e., $\lambda_B^A = \Omega((\frac{N_B}{N})^{1-\frac{\kappa}{2}+\mu})$. Recall that $N_B = N^\nu$, $0 < \nu < 1$. Hence, $\lambda_B^A = \Omega(N^{(\nu-1)(1-\frac{\kappa}{2}+\mu)})$.

As per (4.9) and (4.13), we obtain a feasible downlink throughput $\lambda_B(N, N_B)$ when related network parameters are given. In the following, we show an asymptotic lower bound of λ_B. Since $\lambda_B^P = \Omega(\frac{N_B}{N}\log_2(\frac{N}{N_B}))$ and $\lambda_B^A = \Omega((\frac{N_B}{N})^{1-\frac{\kappa}{2}+\mu})$, we have

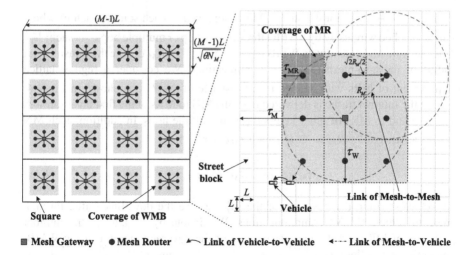

Fig. 4.4 Grid-like VANETs with deployment of WMBs

(i) When $\mu < \frac{\kappa}{2}$, $\lambda_B(N, N_B) = \Omega\left(\frac{N_B}{N} \log_2\left(\frac{N}{N_B}\right)\right)$;

(ii) When $\frac{\kappa}{2} \leq \mu < \kappa$, $\lambda_B(N, N_B) = \Omega\left(\left(\frac{N_B}{N}\right)^{1-\frac{\kappa}{2}+\mu}\right)$.

Notably, the downlink throughput of the network mostly depends on the number of deployed BSs, the coverage of the BS, and the transmission radius of the vehicle. For the case in which the transmission range of vehicles is relatively small compared with the coverage of BSs, the downlink throughput of B2V transmissions is lower than that of V2V transmissions and thereby determines the network throughput; with a relatively large vehicular transmission range, V2V communications limit the network throughput due to that medium access probability of vehicles is quite small and thereby degrades the per-vehicle throughput in V2V transmissions.

4.3.2 Network with Deployment of WMBs

Figure 4.4 shows the network with deployment of WMBs. There are N_M MNs in the network, θN_M of which are functioned as mesh gateways (MGs) connecting to the Internet through the wireline, where $0 < \theta < 1$. Similar to BSs, MGs are regularly deployed in the city grid, each of which is deployed at the center of a square of area $\frac{(M-1)^2 L^2}{\theta N_M}$. Let τ_M denote the number of tiers of each square. We have,

$$\tau_M \leq \left\lceil \frac{M-1}{2\sqrt{\theta N_M}} + 1 \right\rceil. \tag{4.14}$$

There are $\frac{(1-\theta)N_M}{\theta N_M}$ mesh routers (MRs) deployed in each square, each of which can be reached wirelessly by the MG through one hop or multiple hops. As such, $\frac{1-\theta}{\theta}$ MRs and one MG constitute a WMB in each square. Let R_M denote the transmission radius of mesh-to-mesh (M2M) communications. We consider a regular lattice deployment of MRs with nearest nodal distance of $\frac{\sqrt{2}}{2}R_M$, as shown in Fig. 4.4, so that the Internet traffic is delivered from the MG to MRs of the first layer through one hop and to MRs of other layers through multiple hops. In addition, each MN covers an area of $\frac{\sqrt{2}}{2}R_M \times \frac{\sqrt{2}}{2}R_M$ with τ_{MR} tiers, where

$$\tau_{MR} \leq \lceil \sqrt{2}R_M/(4L) + 1 \rceil. \tag{4.15}$$

Vehicles within the coverage of the MN receive the downlink traffic through mesh-to-vehicle (M2V) communications. We denote by Q and τ_W the number of layers of MRs and the number of tiers of the coverage region of each WMB, respectively. It follows that $\sum_{q=1}^{Q-1} 8q \leq (1-\theta)/\theta$. Thus, $Q \leq \frac{1}{2}\sqrt{(1-\theta)/\theta} + 1$. We have

$$\tau_W \leq \lceil \frac{\sqrt{2}R_M(3 + \sqrt{(1-\theta)/\theta})}{4L} \rceil. \tag{4.16}$$

When $\tau_W > \tau_M$, let $\tau_W = \tau_M$. The network is completely covered by WMBs if $\tau_W = \tau_M$, otherwise not. In the case where $\tau_W < \tau_M$, vehicles in the area without the WMB coverage receive the downlink traffic through V2V transmissions and require the assistance of vehicles on road segments of $Tier(\tau_W)$. We denote the downlink capacity for the deployment of WMBs by $\lambda_M(N, N_M)$. Furthermore, we denote by λ_M^M, λ_M^P, and λ_M^A the downlink capacity of M2M, M2V, and V2V transmissions in the hybrid mode, respectively.

We first study λ_M^M for delivering Internet traffic from the MG to MRs. All the MNs apply the same transmission power P_M for M2M transmissions. The total bandwidth W is divided into W_1, W_2, and W_3 respectively for M2M, M2V, and V2V transmissions. It follows that $W = W_1 + W_2 + W_3$. We consider that M2M communications are under the coordination of CSMA scheme with carrier sensing radius $2R_M$. Let I_M denote the interference experienced by a receiver in M2M transmissions. Similar to the derivation of I_{γ_0}, I_M can be upper bounded as follows.

$$I_M \leq \frac{6KP_M}{R_M^\beta}\left(1 + \frac{1}{\sqrt{3}(\beta - 2)(\sqrt{3} - 1)^{\beta-2}}\right).$$

Hence, the SINR of the M2M transmission is given by $SINR_M \geq \mathcal{U}_3(\beta)$. Note that on the average, every MG needs to deliver the downlink traffic for $\frac{1-\theta}{\theta}$ MRs. Given a carrier sensing radius of $2R_M$, an average medium access probability over all MNs, denoted by P'_{ac}, is at least $P'_{ac} = 1/\sum_{q=1}^2 8q$. Especially, $P'_{ac} = 1$ for $Q = 1$ and $P'_{ac} \geq 1/9$ for $Q = 2$. Hence, λ_M^M can be lower bounded as follows.

$$\lambda_M^M \geq \frac{W_1 \log_2(1 + SINR_M) P'_{ac}}{(1-\theta)/\theta}$$

$$\geq \frac{W_1 \log_2(1 + \mathcal{U}_3(\beta)) P'_{ac}}{(1-\theta)/\theta}. \tag{4.17}$$

We then study λ_M^P for delivering traffic from the MN to vehicles within its coverage. Similarly, to mitigate the interference from neighboring MNs in M2V transmissions, an MN and its neighbors (at most eight) use different channels for M2V transmissions, each of which has bandwidth $W_2/9$. Let P_{MV} denote the transmission power for M2V communications. The interference experienced by vehicles in M2V communications, denoted by I_{MV}, is given by

$$I_{MV} \leq \sum_{q=1}^{\infty} \frac{8qKP_{MV}}{\left[(3q - \frac{1}{2})\frac{\sqrt{2}}{2}R_M\right]^\beta} \leq \frac{2^{\frac{3}{2}\beta+1}KP_{MV}}{5^\beta R_M^\beta} \cdot \frac{12\beta + 1}{3\beta - 6}.$$

Let P_{MV}^τ denote the received power of a vehicle on the road segment of $Tier(\tau)$ from its own MN, where $\tau \leq \tau_{MR}$. Since $P_{MV}^\tau \geq KP_{MV}/(\sqrt{2}L(\tau - \frac{1}{2}))^\beta$, we have

$$SINR_\tau' \geq \frac{5^\beta(3\beta - 6)}{(12\beta + 1)2^{2\beta+1}} \left[\frac{R_M}{(\tau - \frac{1}{2})L}\right]^\beta, \tag{4.18}$$

where $SINR_\tau'$ is the SINR of the received signal from the MN for vehicles on road segments of $Tier(\tau)$.

Similar to the deployment of BSs, W_τ out of $W_2/9$ is the bandwidth allocated to a single vehicle on the road segment of $Tier(\tau)$ for each coverage of MNs. Since vehicles on road segments of $Tier(\tau_W)$ of the WMB are required to relay the downlink traffic, additional bandwidth needs to be allocated to vehicles on the road segments of $Tier(\tau_{MR})$ for MNs located in the outmost layer Q of the WMB, as shown in Fig. 4.4. In the following, we consider an MN on the boundary of the WMB and derive a lower bound of λ_M^P. For vehicles of $Tier(\tau)$, where $\tau \leq \tau_{MR} - 1$, we have

$$\lambda_M^P = W_\tau \log_2(1 + SINR_\tau'). \tag{4.19}$$

We denote by $\bar{\eta}_2$ the average number of vehicles that need a vehicle of $Tier(\tau_W)$ to relay the downlink traffic. We have,

$$\bar{\eta}_2 = \frac{\sum_{\tau=\tau_W+1}^{\tau_M} 16\tau - 12}{16\tau_W - 12} \leq \frac{\tau_M^2 - \tau_W^2}{\tau_W - 1}. \tag{4.20}$$

Hence,

$$\lambda_M^P = \frac{W_{\tau_{MR}} \log_2(1 + SINR_{\tau_{MR}}')}{1 + \bar{\eta}_2}. \tag{4.21}$$

From (4.19) to (4.21), it holds that $\lambda_M^P = \frac{W_2/9}{\xi L \mathcal{U}_4}$, where

$$\mathcal{U}_4 = \sum_{\tau=1}^{\tau_{MR}-1} \frac{(16\tau - 12)}{\log_2(1 + SINR_\tau')} + \frac{(16\tau_{MR} - 12)(1 + \bar{\eta}_2)}{\log_2(1 + SINR_{\tau_{MR}}')}$$

$$\leq \frac{4\tau_{MR}(2\tau_{MR} - 1) + \bar{\eta}_2(16\tau_{MR} - 12)}{\log_2(1 + SINR_{\tau_{MR}}')}.$$

Let $\mathcal{U}_5$ denote the numerator of the last fraction, which is an upper bound of the average number of vehicles served by a single MN. From (4.14) to (4.16), we can attain a lower bound of λ_M^P, i.e.,

$$\lambda_M^P \geq \frac{W_2 \log_2(1 + SINR_{\tau_{MR}}')}{9\xi L \mathcal{U}_5}$$

$$\sim \frac{W_2}{9\xi L \mathcal{U}_5} \log_2\left(1 + \frac{5^\beta(3\beta - 6)}{(12\beta + 1)2^{\frac{1}{2}\beta + 1}}\right). \tag{4.22}$$

Furthermore, let $N_M = N^\gamma$, where $0 < \gamma < 1$. We have $\lambda_M^P = \Omega(\frac{N_M}{N}) = \Omega(N^{\gamma-1})$ asymptotically.

We follow the derivation of (4.13) to obtain λ_M^A, since V2V communications are with the same configurations in both BSs and WMBs deployments. Hence,

$$\lambda_M^A \geq \frac{W_3 \log_2(1 + SINR_V)P_{ac}}{\bar{\eta}_2}$$

$$\geq \frac{W_3 \log_2(1 + \mathcal{U}_3(\beta))(\tau_W - 1)}{8\xi L(2R_V/L + 1)^2(\tau_M^2 - \tau_W^2)}. \tag{4.23}$$

Asymptotically,

$$\lambda_M^A = \Omega\left(\frac{N_M(R_M/L)}{N(R_V/L)^2}\right).$$

We let $(R_M/L) = \tau_M^{\sigma_1}$ to establish a relationship between the transmission range of MNs and the size of the mesh square, where $0 < \sigma_1 < 1$. Similarly, $R_V/L = \tau_M^{\sigma_2}$, where $0 < \sigma_2 < 1$ and $\sigma_2 < \sigma_1$. Therefore, $\lambda_M^A = \Omega(N^{(\gamma-1)(1+\sigma_2-\frac{1}{2}\sigma_1)})$. From (4.17), (4.22), and (4.23), we can attain a lower bound of $\lambda_M(N, N_M)$ as follows.

$$\lambda_M(N, N_M) = \min\left(\frac{\lambda_M^M}{\mathcal{U}_5}, \min(\lambda_M^P, \lambda_M^A)\right). \tag{4.24}$$

Notably, $\lambda_M^M/\mathcal{U}_5 = \Omega(N^{\gamma-1})$. Then, we obtain the following asymptotic bound of λ_M^M in the hybrid mode:

(i) When $\sigma_2 < \frac{1}{2}\sigma_1$,

$$\lambda_M(N, N_M) = \Omega\left(\frac{N_M}{N}\right);$$

(ii) When $\frac{1}{2}\sigma_1 \leq \sigma_2 < \sigma_1$,

$$\lambda_M(N, N_M) = \Omega\left(\left(\frac{N_M}{N}\right)^{1-\frac{1}{2}\sigma_1+\sigma_2}\right).$$

When the network is fully covered by deployed WMBs, each MN covers an area of $(M-1)^2L^2/N_M$. Hence, $R_M \geq \sqrt{2}(M-1)L/\sqrt{N_M}$. We then have

$$\lambda_M^P \geq \frac{(W-W_1)\log_2(1+SINR'_{\tau_{MR}})}{9N/N_M}$$

$$\sim \frac{(W-W_1)N_M}{9N}\log_2\left(1+\frac{5^\beta(3\beta-6)}{(12\beta+1)2^{\frac{1}{2}\beta+1}}\right).$$

It follows that $\lambda_M(N, N_M) = \min(N_M\lambda_M^M/N, \lambda_M^P)$ in the infrastructure mode. Asymptotically, $\lambda_M(N, N_M) = \Omega(N_M/N) = \Omega(N^{\gamma-1})$.

4.3.3 Network with Deployment of RAPs

We consider that the coverage of the RAP is one-dimensional along the road, as shown in Fig. 4.5. There are N_R RAPs regularly deployed in the network and each RAP provides Internet access service to vehicles on the road of length L_R, which is called the RAP cell. It can be seen that $L_R = \frac{2(M-1)^2L}{N_R}$. The coverage radius of RAP is denoted by R_C. When $R_C > \frac{1}{2}L_R$, let $R_C = \frac{1}{2}L_R$. The network is fully covered by RAPs if $R_C = \frac{1}{2}L_R$. To provide pervasive Internet access, the network operates in the hybrid mode when $R_V < R_C < \frac{1}{2}L_R$: vehicles within the coverage of the RAP receive the downlink traffic through RAP-to-vehicle (R2V) communications; vehicles at distance $(R_C - R_V, R_C]$ from the RAP are required to relay the downlink traffic for vehicles in the area without the RAP coverage, given the transmission radius of V2V communications R_V. The downlink capacity for the

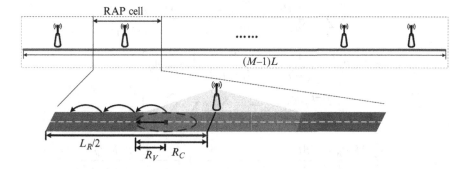

Fig. 4.5 Grid-like VANETs with deployment of RAPs

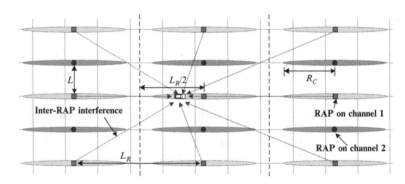

Fig. 4.6 An illustration of inter-RAP interference for horizontal roads

deployment of RAPs is denoted by $\lambda_R(N, N_R)$. Moreover, the downlink capacity of R2V and V2V transmissions are denoted by λ_R^P and λ_R^A, respectively. Similarly, in the hybrid mode,

$$\lambda_R(N, N_R) = \min\{\lambda_R^P, \lambda_R^A\}. \tag{4.25}$$

We first study the downlink throughput λ_R^P in the hybrid mode. A spectrum reuse scheme is adopted to mitigate the inter-RAP interference: (i) RAPs deployed along the same road operate on one common channel; (ii) RAPs on any two adjacent parallel roads use different channels; and (iii) RAPs on the horizontal roads and on the vertical roads use different channels. To that end, four different communication channels, each of which has bandwidth $\frac{1}{4}\phi W$, are allocated. The remaining bandwidth of $(1 - \phi)W$ is allocated for V2V communications. The interference I_d experienced by a vehicle at distance d from the RAP, where $d \leq R_C$, in R2V communications is the aggregated signal power from all the other RAPs operating on the same channel, as shown in the Fig. 4.6. We have

$$I_d \leq \sum_{q=1}^{\infty} \left[\frac{KP_R}{(qL_R - d)^{\beta}} + \frac{KP_R}{(qL_R + d)^{\beta}} \right]$$

$$+ \sum_{q=1}^{\infty} \frac{2KP_R}{(2qL)^{\beta}} + \sum_{i=1}^{\infty} \sum_{j=1}^{\infty} \frac{4KP_R}{(i^2(2L)^2 + j^2 L_R^2)^{\frac{\beta}{2}}}$$

$$\leq 2KP_R \left[\frac{1}{(L_R - d)^{\beta}} + \int_1^{\infty} \frac{1}{(qL_R - d)^{\beta}} dq \right]$$

$$+ \frac{2^{1-\beta}\beta KP_R}{(\beta - 1)L^{\beta}} + \frac{2^{2-\beta}KP_R}{(LL_R)^{\frac{\beta}{2}}} \sum_{i=1}^{\infty} \sum_{j=1}^{\infty} \frac{1}{(ij)^{\frac{\beta}{2}}}$$

$$\leq \frac{2KP_R}{\beta - 1} \left(\frac{\beta L_R - d}{L_R(L_R - d)^{\beta}} + \frac{\beta}{(2L)^{\beta}} \right) + \frac{2^{2-\beta}\beta^2 KP_R}{(\beta - 2)^2(LL_R)^{\frac{\beta}{2}}},$$

where P_R is the transmission power of RAPs. The SINR of received signal from the RAP is thereby given as follows.

$$SINR_d \geq \frac{(\beta - 1)/(2d^\beta)}{\frac{\beta L_R - d}{L_R(L_R - d)^\beta} + \frac{\beta}{(2L)^\beta} + \frac{2^{1-\beta}(\beta-1)\beta^2}{(\beta-2)^2(LL_R)^{\frac{\beta}{2}}}} = \mathscr{U}_6(d).$$

For vehicle $\mathscr{V}_d$ at distance d from the RAP, where $d \leq R_C$, it holds that

$$\lambda_R^P = W_d \log_2(1 + SINR_d)$$

where W_d out of $\frac{1}{4}\phi W$ is the bandwidth allocated to $\mathscr{V}_d$. As discussed before, vehicles at distance $(R_C - R_V, R_C]$ from the RAP are required to relay the downlink traffic to the vehicles at distance $(R_C, \frac{1}{2}L_R]$, which yields an average relaying traffic load of that $\bar{\eta}_3 = (\frac{1}{2}L_R - R_C)/R_V$. Therefore, for vehicles at distance $d \in (R_C - R_V, R_C]$ from the RAP,

$$\lambda_R^P = \frac{W_d \log_2(1 + SINR_d)}{1 + \bar{\eta}_3}.$$

Given the constraint of the total bandwidth, we have

$$\lambda_R^P \geq \frac{\frac{1}{4}\phi W}{\frac{2\xi(R_C - R_V)}{\log_2(1 + SINR_{R_C - R_V})} + \frac{2\xi(1+\bar{\eta}_3)R_V}{\log_2(1 + SINR_{R_C})}}$$

$$\geq \frac{\frac{1}{8}\phi W/\xi}{\frac{R_C - R_V}{\log_2(1 + \mathscr{U}_6(R_C - R_V))} + \frac{R_V + \frac{1}{2}L_R - R_C}{\log_2(1 + \mathscr{U}_6(R_C))}}. \tag{4.26}$$

Further, we let $R_C = (\frac{1}{2}L_R)^{\rho_1}$ and $R_V = (\frac{1}{2}L_R)^{\rho_2}$, where $0 < \rho_2 < \rho_1 < 1$. Denoting $N_R = N^\varphi$, where $0 < \varphi < 1$, it can be obtained that $\lambda_R^P = \Omega(\frac{N_R}{N}\log_2(\frac{N}{N_R})) = \Omega(N^{\varphi-1}\log_2 N)$ asymptotically when $\rho_1 < \frac{1}{2}$; $\lambda_R^P = \Omega(\frac{N_R}{N}) = \Omega(N^{\varphi-1})$ when $\rho_1 = \frac{1}{2}$; $\lambda_R^P = \Omega(\frac{N_R}{N}\log_2(1 + (\frac{N_R}{N})^{\beta(\rho_1 - \frac{1}{2})})) = \Omega(N^{(\varphi-1)[1+\beta(\rho_1 - \frac{1}{2})]})$ when $\rho_1 > \frac{1}{2}$. The derivation of λ_R^A is in the same manner.

$$\lambda_R^A \geq \frac{(1 - \phi)W \log_2(1 + SINR_V)P_{ac}}{\bar{\eta}_3}$$

$$\geq \frac{(1 - \phi)W \log_2(1 + \mathscr{U}_3(\beta))R_V}{8\xi L(2R_V/L + 1)^2(\frac{1}{2}L_R - R_C)}. \tag{4.27}$$

Asymptotically, $\lambda_R^A = \Omega((N_R/N)^{1+\rho_2}) = \Omega(N^{(\varphi-1)(1+\rho_2)})$.

As per (4.26) and (4.27), $\lambda_R(N, N_R)$ can be obtained from (4.25) when values of all the impact factors are determined. In addition, the asymptotic bound of $\lambda_R(N, N_R)$ is given by

Table 4.2 Values of parameters

Parameter	Value	Parameter	Value
M	201	L	100 m
ξ	0.05 veh/m	N	4×10^5
W	10 MHz	β	4
R_V	100 m	θ	0.25

(i) When $\rho_1 \leq \frac{1}{2}$,

$$\lambda_R(N, N_R) = \Omega((N_R/N)^{1+\rho_2});$$

(ii) When $\frac{1}{2} < \rho_1 < 1$,

$$\lambda_M(N, N_M) = \Omega\left((N_R/N)^{\max[1+\rho_2, 1+\beta(\rho_1 - \frac{1}{2})]}\right).$$

Especially, when the network is completely covered by RAPs, $\lambda_R(N, N_R) = \lambda_R^P \geq W N_R \log_2(1 + \mathscr{U}_6(R_C))/(4N)$. The asymptotic result of $\lambda_R(N, N_R)$ in the infrastructure mode is the same as that of λ_R^P in the hybrid mode.

4.4   Case Study

We present a case study of downlink capacity of vehicles based on the analytical results from Sect. 4.3. The objective is to evaluate the impact of key factors, i.e., the number of infrastructure nodes deployed and the coverage of infrastructure nodes, on capacity performance and compare the three types of infrastructures in terms of the deployment cost. The values of parameters for this study are given in Table 4.2.

4.4.1   Impact of Coverage of Infrastructure Nodes

We consider a ideal city grid of 20 × 20 km with an average vehicle density of 0.05 vehicle per meter (veh/m). The total bandwidth of 10 MHz is applied in all 3 types of infrastructure deployment. The bandwidth allocation is done to maximize the downlink throughput for each case. The downlink capacity is plotted with respect to the number of deployed infrastructure nodes, as shown in Figs. 4.7–4.9. With the increase of the number of deployed infrastructure nodes, the network transits from being partially covered to being fully covered, and accordingly the downlink throughput increases gradually. The impact of coverage of infrastructure nodes on downlink throughput is also investigated. Three different sizes of BS footprint are considered in Fig. 4.7. We show that for each BS coverage, the achievable downlink throughput increases faster than a linear increase with N_B in the hybrid mode. The

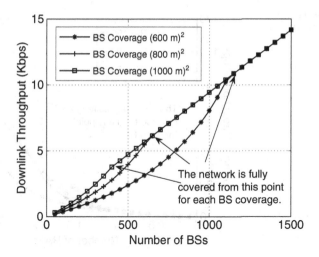

Fig. 4.7 Network with deployment of BSs

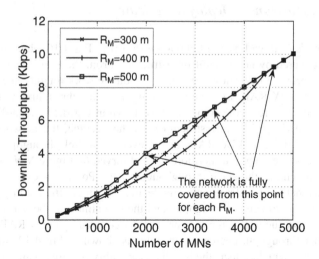

Fig. 4.8 Network with deployment of WMBs

reason is that the relaying traffic load of relay vehicles decreases very fast when the network gradually becomes fully covered and thereby the capacity of V2V communications increases. When the network is fully covered by BSs, the downlink throughput increases almost linearly with N_B. In addition, it is very intuitive that the network needs more BSs to be fully covered with a smaller size of BS coverage. The similar insights for the other two deployments are obtained from Figs. 4.8 and 4.9.

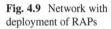

Fig. 4.9 Network with
deployment of RAPs

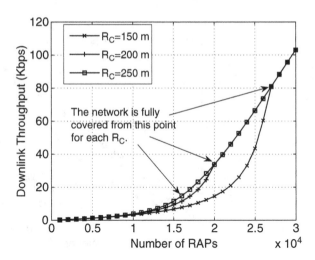

4.4.2   Comparison of Deployment Scales

Different trends of downlink throughput are shown in Fig. 4.10 when the network is not fully covered by any type of infrastructure. From the average slope of each curve, an important observation is attained that the network roughly needs X BSs, or $6X$ MNs, or $25X$ RAPs to achieve a certain downlink throughput in the hybrid mode. A whole picture of the comparison is shown in Fig. 4.11. Regardless of the operation mode (hybrid or infrastructure), on the average, the network requires X BSs, or $5X$ MNs, or $15X$ RAPs to achieve a downlink throughput less than 15 Kbps with our settings. Furthermore, it is observed that more MNs are needed than RAPs to achieve the same throughput after the Point A shown in Fig. 4.11. The reason is that in the infrastructure mode, the relaying traffic load from the MG to MRs limits the downlink throughput, and there is almost no benefit from better coverage of MNs due to that the network is fully covered by either RAPs or MNs. The downlink throughput decreases severely, as shown in Fig. 4.12, with a very small value of θ, which reflects the backhaul capability of wireless mesh networks. Another observation from Fig. 4.11 is that we roughly need to additionally deploy X BSs, or $5X$ MNs, or $1.5X$ RAPs to improve the downlink throughput by the same amount, given that the network operates in the infrastructure mode.

4.4.3   Capacity-Cost Tradeoffs

Deployment cost plays an important role in determining the cost-effective access infrastructure. CAPEX and OPEX contribute to the major component of the deployment cost [138]. As per the cost models in [138], the estimated deployment cost of each type of access infrastructure is described in Table 4.3. It can be seen that when the network operates in the hybrid mode (low-capacity regime), the

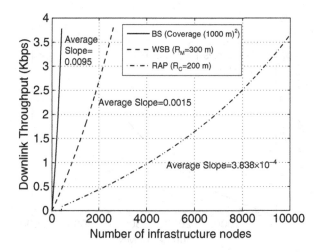

Fig. 4.10 Comparison of number of deployed infrastructure nodes in the hybrid mode

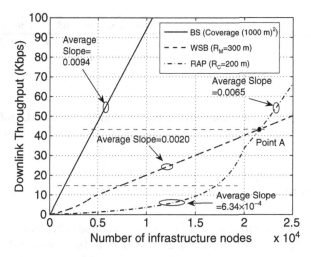

Fig. 4.11 Comparison of number of deployed infrastructure nodes in the infrastructure mode

deployment of BSs or WMBs is cost-effective for a 5-year operation period (the cost is roughly $120X$ K€ to deploy X BSs, or $6X$ MNs). On the other hand, when the network operates in the infrastructure mode (high-capacity regime), the deployment of RAPs outperforms the other two alternatives in terms of deployment costs for a given downlink throughput requirement. For instance, to provide a downlink throughput of 40 Kbps to all the vehicles, roughly we need to pay 530 M€ for the deployment of 4,200 BSs, or 210 M€ for the deployment of 2.1×10^4 RAPs for a 5-year period. From Fig. 4.11, the choice of the cost-effective access infrastructure

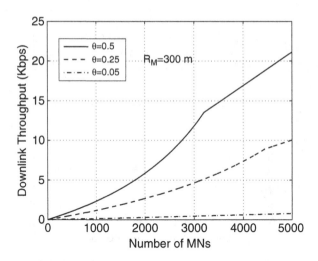

Fig. 4.12 Impact of θ on the downlink throughput for the deployment of WMBs

Table 4.3 Estimated deployment cost (K€)

Deployment cost	BS	MG (MR)	RAP
CAPEX	58.9	10.9 (7.0)	3.0
OPEX (per year)	13.4	2.9 (2.0)	1.4
5-Year cost	125.9	25.4 (17.0)	10.0

can be made as per the data demand of vehicles. Notably, non-cellular infrastructure like RAPs is a good choice to offer a cost-effective high-speed data pipe for vehicles.

4.5 Summary

In this chapter, we have investigated the capacity-cost tradeoffs for different wireless access infrastructures in vehicular access networks. The considered alternatives of wireless access infrastructure include BSs, WMBs, and RAPs, which are respectively deployed to provide downlink Internet service to all the vehicles uniformly in the network. The downlink capacity of vehicles for each kind of deployment has been lower-bounded under the same set of benchmark models by considering a ideal city grid with vehicles randomly distributed on the roads. Moreover, asymptotic results, i.e., in the scaling sense, have also been provided for a large-scale deployment. A case study has been presented to examine the capacity-cost tradeoffs of different solutions in terms of both CAPEX and OPEX. Offering fundamental guidance, results in this chapter imply that it is necessary to choose a cost-effective access infrastructure according to the data demand of vehicles.

Chapter 5
Conclusions and Future Directions

In this chapter, we summarize the main concepts and results presented in this monograph and highlight future research directions.

5.1 Conclusions

In this monograph, we have investigated the capacity of vehicular networks. Based on the analysis and discussion provided throughout this monograph, we present the following remarks.

- Capacity studies play an vital role in understanding the fundamental properties of VANETs. Instead of showing the exact performance for a specific network, theoretical capacity bounds present the performance limit of networks with optimal operations, and thereby guide the network design and deployment. For large-scale VANETs, capacity scaling results can also be applied to predict the network performance, at least in the order sense.
- As vehicular communications are intensively affected by the social behaviors of drivers, we argue that to accurately model the social features of vehicle mobility is crucial for the study of vehicular communications. That is the main implication from the study of social-proximity VANETs, where the capacity highly depends on the inherent mobility patterns of the network. From the capacity and delay analysis, we also know that according to different mobility patterns, it is beneficial to design suitable packet forwarding schemes.
- Wireless access technologies cannot be circumvented when it comes to VANETs, since providing Internet services is the major solution to meet the ever-increasing mobile data demand of vehicle users. The capacity-cost study provides some basic ideas for the choice of access infrastructures. The main conclusion is that the deployment of BSs or WMBs is cost-effective to offer a low-speed downlink rate to vehicles; nonetheless, when providing a high-speed Internet access, the deployment of RAPs outperforms the other two alternatives in terms

N. Lu and X. Shen, *Capacity Analysis of Vehicular Communication Networks*,
SpringerBriefs in Electrical and Computer Engineering,
DOI 10.1007/978-1-4614-8397-7__5, © The Author(s) 2014

of deployment costs. Therefore, we believe that non-cellular access infrastructure will play an increasingly important role in offering a cost-effective data pipe for vehicles, especially for supporting high-bandwidth applications.

5.2 Future Research Directions

In this monograph, we focus on the study of social-proximity VANETs and vehicular access networks, mainly based on the theoretical analysis. Our future work includes extensive simulation validations base on trace data of real-world scenarios and further digging up the implication on network design and operation. Despite existing studies on capacity analysis of VANETs, many issues remain unclear. For example, when jointly considering more complex street patten and inhomogeneous vehicle densities, it might be difficult to determine the throughput capacity and network delay. Moreover, due to the emergent and public nature of safety applications, broadcasting leads an important role in disseminating safety messages to vehicles in proximity. The study of broadcast capacity is another research interest. We close this chapter with additional two thoughts on future research directions in this field.

- The design, analysis and deployment of VANETs necessitate a general understanding of capacity scaling laws. Existing works often adopt different methodologies and sets of assumptions and models in developing capacity scaling laws, which may yield custom-designed solutions without universal properties that can be applied to other scenarios. To better understand the impact of various settings and techniques on capacity scaling laws, it would be useful to provide a unified framework. Two works have been done for general wireless networks toward this end: the study of capacity scaling laws under a generalized physical model [139] and the establishment of a simple set of criteria that can be used to determine the capacity for various physical layer technologies under the protocol model [140]. Nonetheless, great efforts are still needed for VANETs.
- The Shannon capacity was achieved by considering arbitrarily delay and vanishingly small error probability. In [19], Andrews et al. referred to a throughput-delay-reliability (TDR) triplet, since these quantities are interrelated. Thus, the throughput capacity would likely be constrained by these two fundamental quantities—delay and reliability jointly. Actually, the link reliability has been considered in studies of transmission capacity [141–143] which is the spatial intensity of attempted transmissions under a target outage of wireless links. For VANETs, safety applications have QoS requirements for both throughput, delay, and reliability. Hence, the tradeoff between throughput capacity, delay and reliability should be investigated, however it will bring much more challenges.

References

1. D. Jiang, V. Taliwal, A. Meier, W. Holfelder, R. Herrtwich, Design of 5.9 GHz DSRC-based vehicular safety communication. IEEE Wirel. Commun. **13**(5), 36–43 (2006)
2. F. Martinez, C. Toh, J. Cano, C. Calafate, P. Manzoni, Emergency services in future intelligent transportation systems based on vehicular communication networks. IEEE Intell. Transp. Syst. Mag. **2**(2), 6–20 (2010)
3. F. Bai, B. Krishnamachari, Exploiting the wisdom of the crowd: localized, distributed information-centric VANETs. IEEE Commun. Mag. **48**(5), 138–146 (2010)
4. R. Lu, X. Lin, T. Luan, X. Liang, X. Shen, Pseudonym changing at social spots: an effective strategy for location privacy in VANETs. IEEE Trans. Veh. Technol. **61**(1), 86–96 (2012)
5. N. Lu, N. Cheng, N. Zhang, X. Shen, J.W. Mark, VeMail: a message handling system towards efficient transportation management, in *Proceedings of IEEE WCNC*, Shanghai, Apr 2013
6. J. Lin, S. Chen, Y. Shih, S. Chen, A study on remote on-line diagnostic system for vehicles by integrating the technology of OBD, GPS, and 3G. World Acad. Sci. Eng. Technol. **56**, 56 (2009)
7. M. Ramadan, M. Al-Khedher, S. Al-Kheder, Intelligent anti-theft and tracking system for automobiles. Int. J. Mach. Learn. Comput. **2**(1), 88–92 (2012)
8. KPMG's global automotive executive survey, [Online].Available: http://www.kpmg.com/GE/en/IssuesAndInsights/ArticlesPublications/Documents/Global-automotive-executive-survey-2012.pdf
9. B. Chen, M. Chan, Mobtorrent: a framework for mobile internet access from vehicles, in *Proceedings of IEEE INFOCOM*, Rio de Janeiro, Apr 2009
10. T. Luan, X. Ling, X. Shen, MAC in motion: impact of mobility on the MAC of drive-thru internet. IEEE Trans. Mobile Comput. **11**(2), 305–319 (2011)
11. V. Bychkovsky, B. Hull, A. Miu, H. Balakrishnan, S. Madden, A measurement study of vehicular internet access using in situ wi-fi networks, in *Proceedings of ACM MobiCom*, Los Angeles, 2006
12. A. Mahajan, N. Potnis, K. Gopalan, A. Wang, Modeling VANET deployment in urban settings, in *Proceedings of ACM MSWiM*, Chania, 2007, pp. 151–158
13. T. Luan, L. Cai, J. Chen, X. Shen, F. Bai, VTube: towards the media rich city life with autonomous vehicular content distribution, in *Proceedings of IEEE SECON*, Salt Lake City, June 2011
14. G. Karagiannis, O. Altintas, E. Ekici, G. Heijenk, B. Jarupan, K. Lin, T. Weil, Vehicular networking: a survey and tutorial on requirements, architectures, challenges, standards and solutions. IEEE Commun. Surv. Tutor. **13**(4), 1–33 (2011)
15. J. Kenney, Dedicated short-range communications (DSRC) standards in the united states. Proc. IEEE **99**(7), 1162–1182 (2011)

N. Lu and X. Shen, *Capacity Analysis of Vehicular Communication Networks*,   77
SpringerBriefs in Electrical and Computer Engineering,
DOI 10.1007/978-1-4614-8397-7, © The Author(s) 2014

16. H. Hartenstein, K. Laberteaux, I. Ebrary, *VANET: Vehicular Applications and Inter-networking Technologies* (Wiley Online Library, Chichester, 2010)

17. H. Hartenstein, K. Laberteaux, A tutorial survey on vehicular ad hoc networks. IEEE Commun. Mag. **46**(6), 164–171 (2008)

18. A. El Gamal, Y. Kim, *Network Information Theory* (Cambridge University Press, Cambridge/New York, 2011)

19. J. Andrews, S. Shakkottai, R. Heath, N. Jindal, M. Haenggi, R. Berry, D. Guo, M. Neely, S. Weber, S. Jafar et al., Rethinking information theory for mobile ad hoc networks. IEEE Commun. Mag. **46**(12), 94–101 (2008)

20. C. Shannon, A mathematical theory of communication. ACM SIGMOBILE Mobile Comput. Commun. Rev. **5**(1), 3–55 (2001)

21. T. Cover, A. Gamal, Capacity theorems for the relay channel. IEEE Trans. Inf. Theory **25**(5), 572–584 (1979)

22. P. Gupta, P. Kumar, The capacity of wireless networks. IEEE Trans. Inf. Theory **46**(2), 388–404 (2000)

23. A. Goldsmith, M. Effros, R. Koetter, M. Médard, A. Ozdaglar, L. Zheng, Beyond shannon: the quest for fundamental performance limits of wireless ad hoc networks. IEEE Commun. Mag. **49**(5), 195–205 (2011)

24. P. Li, M. Pan, Y. Fang, The capacity of three-dimensional wireless ad hoc networks, in *Proceedings of IEEE INFOCOM*, Shanghai, Apr 2011

25. H. Pishro-Nik, A. Ganz, D. Ni, The capacity of vehicular ad hoc networks, in *Proceedings of Allerton Conference*, University of Illinois at Urbana-Champaign, Monticello, Illinois, USA 2007

26. M. Nekoui, A. Eslami, H. Pishro-Nik, Scaling laws for distance limited communications in vehicular ad hoc networks, in *Proceedings of IEEE ICC*, Beijing, 2008, pp. 2253–2257

27. G. Zhang, Y. Xu, X. Wang, X. Tian, J. Liu, X. Gan, H. Yu, L. Qian, Multicast capacity for hybrid VANETs with directional antenna and delay constraint. IEEE J. Sel. Areas Commun. **30**(4), 818–833 (2012)

28. M. Wang, H. Shan, L. Cai, N. Lu, X. Shen, F. Bai, Throughput capacity of VANETs by exploiting mobility diversity, in *Proceedings of IEEE ICC*, Ottawa, June 2012

29. N. Lu, T. Luan, M. Wang, X. Shen, F. Bai, Bounds of asymptotic performance limits of social-proximity vehicular networks. IEEE/ACM Trans. Netw. to appear

30. N. Lu, N. Zhang, N. Cheng, X. Shen, J.W. Mark, F. Bai, Vehicles meet infrastructure: towards capacity-cost tradeoffs for vehicular access networks. IEEE Trans. Intell. Transp. Syst. to appear

31. T. Cover, J. Thomas, J. Wiley et al., *Elements of Information Theory*, vol. 306 (Wiley Online Library, John Wiley & Sons, Inc., Hoboken, New Jersey 1991)

32. M. Franceschetti, O. Dousse, D. Tse, P. Thiran, Closing the gap in the capacity of wireless networks via percolation theory. IEEE Trans. Inf. Theory **53**(3), 1009–1018 (2007)

33. S. Yi, Y. Pei, S. Kalyanaraman, On the capacity improvement of ad hoc wireless networks using directional antennas, in *Proceedings of ACM MobiHoc*, Annapolis, 2003

34. C. Peraki, S. Servetto, On the maximum stable throughput problem in random networks with directional antennas, in *Proceedings of ACM MobiHoc*, Annapolis, 2003

35. H. Sadjadpour, Z. Wang, J. Garcia-Luna-Aceves, The capacity of wireless ad hoc networks with multi-packet reception. IEEE Trans. Commun. **58**(2), 600–610 (2010)

36. J. Garcia-Luna-Aceves, H. Sadjadpour, Z. Wang, Challenges: towards truly scalable ad hoc networks, in *Proceedings of ACM MobiCom*, Montreal, 2007

37. Z. Wang, H. Sadjadpour, J. Garcia-Luna-Aceves, The capacity and energy efficiency of wireless ad hoc networks with multi-packet reception, in *Proceedings of MobiHoc*, Hong Kong. (ACM, 2008), pp. 179–188

38. Z. Wang, H. Sadjadpour, J. Garcia-Luna-Aceves, Fundamental limits of information dissemination in wireless ad hoc networks–part II: multi-packet reception. IEEE Trans. Wirel. Commun. **10**(3), 803–813 (2011)

39. S. Aeron, V. Saligrama, Wireless ad hoc networks: strategies and scaling laws for the fixed SNR regime. IEEE Trans. Inf. Theory 53(6), 2044–2059 (2007)
40. A. Ozgur, O. Lévêque, D. Tse, Hierarchical cooperation achieves optimal capacity scaling in ad hoc networks. IEEE Trans. Inf. Theory 53(10), 3549–3572 (2007)
41. J. Ghaderi, L. Xie, X. Shen, Hierarchical cooperation in ad hoc networks: optimal clustering and achievable throughput. IEEE Trans. Inf. Theory 55(8), 3425–3436 (2009)
42. U. Niesen, P. Gupta, D. Shah, On capacity scaling in arbitrary wireless networks. IEEE Trans. Inf. Theory 55(9), 3959–3982 (2009)
43. M. Franceschetti, M. Migliore, P. Minero, The capacity of wireless networks: information-theoretic and physical limits. IEEE Trans. Inf. Theory 55(8), 3413–3424 (2009)
44. S. Lee, S. Chung, Capacity scaling of wireless ad hoc networks: Shannon meets Maxwell. IEEE Trans. Inf. Theory 58(3), 1702–1715 (2012)
45. K. Lu, S. Fu, Y. Qian, H. Chen, On capacity of random wireless networks with physical-layer network coding. IEEE J. Sel. Areas Commun. 27(5), 763–772 (2009)
46. J. Liu, D. Goeckel, D. Towsley, The throughput order of ad hoc networks employing network coding and broadcasting, in *IEEE Proceedings of MILCOM*, Washington, DC, Oct 2006
47. J. Liu, D. Goeckel, D. Towsley, Bounds on the gain of network coding and broadcasting in wireless networks, in *IEEE Proceedings of INFOCOM*, Anchorage, May 2007
48. A. Keshavarz-Haddadt, R. Riedi, Bounds on the benefit of network coding: throughput and energy saving in wireless networks, in *IEEE Proceedings of INFOCOM*, Phoenix, Mar 2008
49. J. Liu, D. Goeckel, D. Towsley, Bounds on the throughput gain of network coding in unicast and multicast wireless networks. IEEE J. Sel. Areas Commun. 27(5), 582–592 (2009)
50. R. Negi, A. Rajeswaran, Capacity of power constrained ad-hoc networks, in *IEEE Proceedings of INFOCOM*, Hong Kong, Mar 2004
51. X. Tang, Y. Hua, Capacity of ultra-wideband power-constrained ad hoc networks. IEEE Trans. Inf. Theory 54(2), 916–920 (2008)
52. H. Zhang, J. Hou, Capacity of wireless ad-hoc networks under ultra wide band with power constraint, in *IEEE Proceedings of INFOCOM*, Miami, Mar 2005
53. M. Grossglauser, D. Tse, Mobility increases the capacity of ad hoc wireless networks. IEEE/ACM Trans. Netw. 10(4), 477–486 (2002)
54. S. Diggavi, M. Grossglauser, D. Tse, Even one-dimensional mobility increases the capacity of wireless networks. IEEE Trans. Inf. Theory 51(11), 3947–3954 (2005)
55. S. Jafar, Too much mobility limits the capacity of wireless ad hoc networks. IEEE Trans. Inf. Theory 51(11), 3954–3965 (2005)
56. P. Kyasanur, N. Vaidya, Capacity of multi-channel wireless networks: impact of number of channels and interfaces, in *Proceedings of ACM MobiCom*, Cologne, 2005
57. P. Kyasanur, N. Vaidya, Capacity of multichannel wireless networks under the protocol model. IEEE/ACM Trans. Netw. 17(2), 515–527 (2009)
58. M. Kodialam, T. Nandagopal, Characterizing the capacity region in multi-radio multi-channel wireless mesh networks, in *Proceedings of ACM MobiCom*, Cologne, 2005
59. S. Toumpis, A. Goldsmith, Large wireless networks under fading, mobility, and delay constraints, in *Proceeding of IEEE INFOCOM*, Hongkong, Mar 2004
60. M. Ebrahimi, M. Maddah-Ali, A. Khandani, Throughput scaling laws for wireless networks with fading channels. IEEE Trans. Inf. Theory 53(11), 4250–4254 (2007)
61. R. Gowaikar, B. Hochwald, B. Hassibi, Communication over a wireless network with random connections. IEEE Trans. Inf. Theory 52(7), 2857–2871 (2006)
62. S. Cui, A. Haimovich, O. Somekh, H. Poor, S. Shamai, Throughput scaling of wireless networks with random connections. IEEE Trans. Inf. Theory 56(8), 3793–3806 (2010)
63. R. Gowaikar, B. Hassibi, Achievable throughput in two-scale wireless networks. IEEE J. Sel. Areas Commun. 27(7), 1169–1179 (2009)
64. R. Jaber, J. Andrews, A lower bound on the capacity of wireless erasure networks. IEEE Trans. Inf. Theory 57(10), 6502–6513 (2011)
65. C. Hu, X. Wang, Z. Yang, J. Zhang, Y. Xu, X. Gao, A geometry study on the capacity of wireless networks via percolation. IEEE Trans. Commun. 58(10), 2916–2925 (2010)

66. P. Li, M. Pan, Y. Fang, Capacity bounds of three-dimensional wireless ad hoc networks. IEEE/ACM Trans. Netw. **20**(4), 1304–1315 (2012)

67. A. Keshavarz-Haddad, V. Ribeiro, R. Riedi, Broadcast capacity in multihop wireless networks, in *Proceedings of ACM MobiCom*, Los Angeles, 2006

68. R. Zheng, Asymptotic bounds of information dissemination in power-constrained wireless networks. IEEE Trans. Wirel. Commun. **7**(1), 251–259 (2008)

69. X. Li, J. Zhao, Y. Wu, S. Tang, X. Xu, X. Mao, Broadcast capacity for wireless ad hoc networks, in *Proceedings of IEEE MASS*, Atlanta, Sept 2008

70. S. Li, Y. Liu, X. Li, Capacity of large scale wireless networks under gaussian channel model, in *Proceedings of ACM MobiCom*, San Francisco, 2008

71. X. Li, Multicast capacity of wireless ad hoc networks. IEEE/ACM Trans. Netw. **17**(3), 950–961 (2009)

72. C. Wang, X. Li, C. Jiang, S. Tang, Y. Liu, J. Zhao, Scaling laws on multicast capacity of large scale wireless networks, in *Proceedings of IEEE INFOCOM*, Rio de Janeiro, Apr 2009

73. S. Shakkottai, X. Liu, R. Srikant, The multicast capacity of large multihop wireless networks. IEEE/ACM Trans. Netw. **18**(6), 1691–1700 (2010)

74. U. Niesen, P. Gupta, D. Shah, The balanced unicast and multicast capacity regions of large wireless networks. IEEE Trans. Inf. Theory **56**(5), 2249–2271 (2010)

75. X. Wang, W. Huang, S. Wang, J. Zhang, C. Hu, Delay and capacity tradeoff analysis for motioncast. IEEE/ACM Trans. Netw. **19**(5), 1354–1367 (2011)

76. X. Wang, L. Fu, C. Hu, Multicast performance with hierarchical cooperation. IEEE/ACM Trans. Netw. **20**(3), 917–930 (2012)

77. D. Nie, A survey on multicast capacity of wireless ad hoc networks (2009). [Online]. Available: http://iwct.sjtu.edu.cn/personal/xwang8/research/nieding/survey.pdf

78. Z. Wang, H. Sadjadpour, J. Garcia-Luna-Aceves, A unifying perspective on the capacity of wireless ad hoc networks, in *Proceedings of IEEE INFOCOM*, Phoenix, Apr 2008

79. G. Sharma, R. Mazumdar, N. Shroff, Delay and capacity trade-offs in mobile ad hoc networks: a global perspective. IEEE/ACM Trans. Netw. **15**(5), 981–992 (2007)

80. D. Ciullo, V. Martina, M. Garetto, E. Leonardi, Impact of correlated mobility on delay-throughput performance in mobile ad-hoc networks, in *Proceedings of IEEE INFOCOM*, San Diego, Mar 2010

81. S. Ross, *Introduction to Probability Models* (Academic, Academic Press, Burlington, MA, USA 2009)

82. K. Lee, Y. Kim, S. Chong, I. Rhee, Y. Yi, Delay-capacity tradeoffs for mobile networks with Lévy walks and Lévy flights, in *Proceedings of IEEE INFOCOM*, San Diego, Mar 2010

83. I. Rhee, M. Shin, S. Hong, K. Lee, S. Kim, S. Chong, On the Levy-walk nature of human mobility. IEEE/ACM Trans. Netw. **19**(3), 630–643 (2011)

84. M. Neely, E. Modiano, Capacity and delay tradeoffs for ad hoc mobile networks. IEEE Trans. Inf. Theory **51**(6), 1917–1937 (2005)

85. L. Ying, S. Yang, R. Srikant, Optimal delay–throughput tradeoffs in mobile ad hoc networks. IEEE Trans. Inf. Theory **54**(9), 4119–4143 (2008)

86. X. Lin, N. Shroff, The fundamental capacity-delay tradeoff in large mobile ad hoc networks, in *Proceedings of 3rd Annual Mediterranean Ad Hoc Networking Workshop*, Bodrum, June 2004

87. A. El Gamal, J. Mammen, B. Prabhakar, D. Shah, Optimal throughput-delay scaling in wireless networks-part I: the fluid model. IEEE Trans. Inf. Theory **52**(6), 2568–2592 (2006)

88. X. Lin, G. Sharma, R. Mazumdar, N. Shroff, Degenerate delay-capacity tradeoffs in ad-hoc networks with brownian mobility. IEEE Trans. Inf. Theory **52**(6), 2777–2784 (2006)

89. G. Sharma, R. Mazumdar, Scaling laws for capacity and delay in wireless ad hoc networks with random mobility, in *Proceedings of IEEE ICC*, Paris, June 2004

90. P. Li, Y. Fang, J. Li, Throughput, delay, and mobility in wireless ad hoc networks, in *Proceedings of IEEE INFOCOM*, San Diego, Mar 2010

91. M. Garetto, P. Giaccone, E. Leonardi, On the capacity of ad hoc wireless networks under general node mobility, in *Proceedings of IEEE INFOCOM*, Anchorage, May 2007

92. M. Garetto, P. Giaccone, E. Leonardi, Capacity scaling of sparse mobile ad hoc networks, in *Proceedings of IEEE INFOCOM*, Phoenix, Apr 2008

93. M. Garetto, P. Giaccone, E. Leonardi, Capacity scaling in ad hoc networks with heterogeneous mobile nodes: the super-critical regime. IEEE/ACM Trans. Netw. **17**(5), 1522–1535 (2009)

94. M. Garetto, P. Giaccone, E. Leonardi, Capacity scaling in ad hoc networks with heterogeneous mobile nodes: the subcritical regime. IEEE/ACM Trans. Netw. **17**(6), 1888–1901 (2009)

95. M. Garetto, E. Leonardi, Restricted mobility improves delay-throughput tradeoffs in mobile ad hoc networks. IEEE Trans. Inf. Theory **56**(10), 5016–5029 (2010)

96. A. Ozgur, O. Lévêque, Throughput-delay tradeoff for hierarchical cooperation in ad hoc wireless networks. IEEE Trans. Inf. Theory **56**(3), 1369–1377 (2010)

97. C. Comaniciu, H. Poor, On the capacity of mobile ad hoc networks with delay constraints. IEEE Trans. Wirel. Commun. **5**(8), 2061–2071 (2006)

98. B. Liu, Z. Liu, D. Towsley, On the capacity of hybrid wireless networks, in *Proceedings of IEEE INFOCOM*, San Francisco, Mar 2003

99. U. Kozat, L. Tassiulas, Throughput capacity of random ad hoc networks with infrastructure support, in *Proceedings of ACM MobiCom*, San Diego, 2003, pp. 55–65

100. S. Toumpis, Capacity bounds for three classes of wireless networks: asymmetric, cluster, and hybrid, in *Proceedings of ACM MobiHoc*, Tokyo, 2004, pp. 133–144

101. A. Zemlianov, G. De Veciana, Capacity of ad hoc wireless networks with infrastructure support. IEEE J. Sel. Areas Commun. **23**(3), 657–667 (2005)

102. B. Liu, P. Thiran, D. Towsley, Capacity of a wireless ad hoc network with infrastructure, in *Proceedings of the 8th ACM International Symposium on mobile Ad Hoc Networking and Computing*, Montréal (ACM, 2007), pp. 239–246

103. P. Li, Y. Fang, Impacts of topology and traffic pattern on capacity of hybrid wireless networks. IEEE Trans. Mobile Comput. **8**(12), 1585–1595 (2009)

104. D. Shila, Y. Cheng, T. Anjali, Throughput and delay analysis of hybrid wireless networks with multi-hop uplinks, in *Proceedings of IEEE INFOCOM*, Shanghai, Apr 2011

105. P. Li, C. Zhang, Y. Fang, Capacity and delay of hybrid wireless broadband access networks. IEEE J. Sel. Areas Commun. **27**(2), 117–125 (2009)

106. G. Zhang, Y. Xu, X. Wang, M. Guizani, Capacity of hybrid wireless networks with directional antenna and delay constraint. IEEE Trans. Commun. **58**(7), 2097–2106 (2010)

107. W. Shin, S. Jeon, N. Devroye, M. Vu, S. Chung, Y. Lee, V. Tarokh, Improved capacity scaling in wireless networks with infrastructure. IEEE Trans. Inf. Theory **57**(8), 5088–5102 (2011)

108. W. Huang, X. Wang, Q. Zhang, Capacity scaling in mobile wireless ad hoc network with infrastructure support, in *Proceedings of IEEE ICDCS*, Genoa, June 2010

109. P. Li, Y. Fang, The capacity of heterogeneous wireless networks, in *Proceedings of IEEE INFOCOM*, San Diego, Mar 2010

110. P. Zhou, X. Wang, R. Rao, Asymptotic capacity of infrastructure wireless mesh networks. IEEE Trans. Mobile Comput. **7**(8), 1011–1024 (2008)

111. L. Law, K. Pelechrinis, S. Krishnamurthy, M. Faloutsos, Downlink capacity of hybrid cellular ad hoc networks. IEEE/ACM Trans. Netw. **18**(1), 243–256 (2010)

112. P. Li, X. Huang, Y. Fang, Capacity scaling of multihop cellular networks, in *Proceedings of IEEE INFOCOM*, Shanghai, Apr 2011

113. N. Zhang, N. Lu, R. Lu, J.W. Mark, X. Shen, Energy-efficient and trust-aware cooperation in cognitive radio networks, in *Proceedings of IEEE ICC*, Ottawa, June 2012

114. M. Vu, V. Tarokh, Scaling laws of single-hop cognitive networks. IEEE Trans. Wirel. Commun. **8**(8), 4089–4097 (2009)

115. S. Jeon, N. Devroye, M. Vu, S. Chung, V. Tarokh, Cognitive networks achieve throughput scaling of a homogeneous network. IEEE Trans. Inf. Theory **57**(8), 5103–5115 (2011)

116. C. Yin, L. Gao, S. Cui, Scaling laws for overlaid wireless networks: a cognitive radio network versus a primary network. IEEE/ACM Trans. Netw. **18**(4), 1317–1329 (2010)

117. W. Huang, X. Wang, Capacity scaling of general cognitive networks. IEEE/ACM Trans. Netw. to appear

118. Y. Li, X. Wang, X. Tian, X. Liu, Scaling laws for cognitive radio network with heterogeneous mobile secondary users, in *Proceedings of IEEE INFOCOM*, Orlando, Mar 2012
119. N. Sarafijanovic-Djukic, M. Pidrkowski, M. Grossglauser, Island hopping: efficient mobility-assisted forwarding in partitioned networks, in *Proceedings of IEEE SECON*, Reston, Sept 2006
120. S. Kostof, R. Tobias, *The City Shaped* (Thames and Hudson, London, 1991)
121. A. Siksna, The effects of block size and form in North American and Australian city centres. Urban Morphol. **1**, 19–33 (1997)
122. M. Neely, E. Modiano, C. Rohrs, Dynamic power allocation and routing for time-varying wireless networks. IEEE J. Sel. Areas Commun. **23**(1), 89–103 (2005)
123. R. Urgaonkar, M. Neely, Network capacity region and minimum energy function for a delay-tolerant mobile ad hoc network. IEEE/ACM Trans. Netw. **19**(4), 1137–1150 (2011)
124. D. Slaughter, *Difference Equations to Differential Equations* (University Press of Florida, Gainesville, FL, USA 2009)
125. R. Motwani, P. Raghavan, *Randomized Algorithms* (Chapman & Hall/CRC, Cambridge University Press, Cambridge, UK 1995)
126. V. Vapnik, A. Chervonenkis, On the uniform convergence of relative frequencies of events to their probabilities. Theory Probab. Appl. **16**, 264 (1971)
127. V. Vapnik, *Statistical Learning Theory* (Wiley-Interscience, New York, 1998)
128. N. Banerjee, M. Corner, D. Towsley, B. Levine, Relays, base stations, and meshes: enhancing mobile networks with infrastructure, in *Proceedings of ACM MobiCom*, San Francisco, 2008
129. J. Eriksson, H. Balakrishnan, S. Madden, Cabernet: vehicular content delivery using WiFi, in *Proceedings of ACM MobiCom*, San Francisco, 2008
130. F. Malandrino, C. Casetti, C.-F. Chiasserini, M. Fiore, Content downloading in vehicular networks: what really matters, in *Proceedings of IEEE INFOCOM*, Shanghai, Apr 2011
131. C. Stefanović, D. Vukobratović, F. Chiti, L. Niccolai, V. Crnojević, R. Fantacci, Urban infrastructure-to-vehicle traffic data dissemination using uep rateless codes. IEEE J. Sel. Areas Commun. **29**(1), 94–102 (2011)
132. I.W.-H. Ho, K.K. Leung, J.W. Polak, Stochastic model and connectivity dynamics for vanets in signalized road systems. IEEE/ACM Trans. Netw. **19**(1), 195–208 (2011)
133. H. Xia, A simplified analytical model for predicting path loss in urban and suburban environments. IEEE Trans. Veh. Technol. **46**(4), 1040–1046 (1997)
134. J. Lee, R. Mazumdar, N. Shroff, Joint resource allocation and base-station assignment for the downlink in CDMA networks. IEEE/ACM Trans. Netw. **14**(1), 1–14 (2006)
135. F. Baccelli, B. Blaszczyszyn, *Stochastic Geometry and Wireless Networks Volume I: Theory*. Foundations and Trends in Networking (NOW, Now Publishers, Hanover 2010)
136. F. Baccelli, B. Blaszczyszyn, *Stochastic Geometry and Wireless Networks Volume II: Applications*. Foundations and Trends in Networking (NOW, Now Publishers, Hanover 2010)
137. M. Haenggi, R. Ganti, *Interference in Large Wireless Networks* (Now Publishers, Hanover, 2009)
138. K. Johansson, Cost effective deployment strategies for heterogeneous wireless networks. Ph.D. Dissertation, Kommunikationsteknik, Kungliga Tekniska högskolan, 2007
139. C. Wang, C. Jiang, X. Li, S. Tang, P. Yang, General capacity scaling of wireless networks, in *Proceedings of IEEE INFOCOM*, Shanghai, Apr 2011
140. C. Jiang, Y. Shi, Y. Hou, W. Lou, S. Kompella, S. Midkiff, Toward simple criteria to establish capacity scaling laws for wireless networks, in *Proceedings IEEE INFOCOM, 2012*, Orlando (IEEE, 2012), pp. 774–782
141. S. Weber, X. Yang, J. Andrews, G. De Veciana, Transmission capacity of wireless ad hoc networks with outage constraints. IEEE Trans. Inf. Theory **51**(12), 4091–4102 (2005)
142. A. Hunter, J. Andrews, S. Weber, Transmission capacity of ad hoc networks with spatial diversity. IEEE Trans. Wirel. Commun. **7**(12), 5058–5071 (2008)
143. S. Weber, J. Andrews, N. Jindal, An overview of the transmission capacity of wireless networks. IEEE Trans. Commun. **58**(12), 3593–3604 (2010)